Student Manual for
Essential Mathematics

A Student Oriented Teaching or Self-Study Text

Second Edition

Rudolf A. Zimmer

Fanshawe College

 KENDALL/HUNT PUBLISHING COMPANY
4050 Westmark Drive   Dubuque, Iowa 52002

Part

Solutions to Drill Exercises and Answers to odd-numbered Assignment Questions

Unit 1

Solutions to Drill Exercises

I   1. 32 406 means 3 ten thousands + 2 thousands + 4 hundreds + 0 tens + 6 units.

2. 3 500 601 means 3 millions + 5 hundred thousands + 0 ten thousands + 0 thousands + 6 hundreds + 0 tens + 1 unit.

0 thousands + 6 hundreds + 0 tens + 1 unit.

II   3.
```
  639
  401
   86
+277
 1403
```

4.
```
  703
  −68
  635
```

5.   3928 x 47
```
=   3928
    x 47
   27496
   15712
  184616
```

6. 4230 ÷ 38
```
         111
     38)4230
         38
         43
         38
         50
         38
         12
```
Hence, 4230 ÷ 38 = 111 R12

7. 5617 x 308
```
=      5617
       x 308
      44936
     16851
    1730036
```

8.
```
  6500
 −4437
  2063
```

9. 22 734 ÷ 206
```
           110
    206)22734
        206
        213
        206
         74
```

Hence, 22 734 ÷ 206 = 110 R74

10.
```
  6194
   708
    26
   613
  7541
```

III   11. $\frac{0}{72}$ is undefined is false; $\frac{0}{72} = 0$

12. 24 327 x (7934 x 10) = (24 327 x 7934) x 10 is true; associative property of multiplication.

13. 486 + 7352 = 7352 + 486 is true; commutative property of addition.

14. 3984 − 2765 = 2765 − 3984 is false; subtraction is *not* commutative.

15. 216 x 1 = 216 is true; identity element of multiplication.

IV   16. $3(4 + 10)$ = $3(4) + 3(10)$ 17. $7(3 + 6)$ = $7(3) + 7(6)$

                  = $12 + 30$ = $21 + 42$

                  = 42 = 63

V   18. $280 + 460 + 80 = 820$

      This is 20 kg over the maximum safe load of 800 kg, so it would not be safe.

    19.   Distance travelled first day: $28\ 039 - 27\ 503 = 536$

      Distance travelled second day: $28\ 514 - 28\ 039 = 475$

      $536 - 475 = 61$

      The distance travelled during first day was greater by 61 km.

    20. $45\ 900 \div 225 = 204$

      Each shareholder received $204.00.

Answers to Assignments

I   1.   8911 means 8 thousands + 9 hundreds + 1 ten + 1 unit

II   3.   3695   5.   107 R5

    7.   4 808 724   9.   89 R28

III 11.   True; distributive property of multiplication with respect to addition.

    13.   True; multiplication by zero.

    15.   True; associative property of addition.

IV   17. $9(3 + 6) = (9 \times 3) + (9 \times 6) = 27 + 54 = 81$ V   19.   8448 seats

---

Unit 2

Solutions to Drill Exercises

I   1. 7^3 2. 5^8 3. $2^3 \times 3^2 \times 7^4$ 4. $3^4 \times 11^2 \times 16^3$

II

	Base	Exponent	Base	Exponent	Base	Exponent	Value
5.	10	3	–	–	–	–	1000
6.	5	4	–	–	–	–	625
7.	2	4	3	3	–	–	432
8.	2	2	5	3	10	2	50 000

III 9. $\sqrt{16} = 4$ 10. $\sqrt{121} = 11$

    11. $\sqrt{14^2} = 14$ 12. $\sqrt{0} = 0$

IV   13.

$$\begin{array}{r|r} 2 & 90 \\ \hline 3 & 45 \\ \hline 3 & 15 \\ \hline 5 & 5 \\ \hline & 1 \end{array}$$

$$\therefore\ 90\ =\ 2 \cdot 3^2 \cdot 5$$

14.

$$\begin{array}{r|r} 2 & 360 \\ \hline 2 & 180 \\ \hline 2 & 90 \\ \hline 3 & 45 \\ \hline 3 & 15 \\ \hline 5 & 5 \\ \hline & 1 \end{array}$$

$$\therefore\ 360\ =\ 2^3 \cdot 3^2 \cdot 5$$

15.

$$\begin{array}{r|r} 3 & 297 \\ \hline 3 & 99 \\ \hline 3 & 33 \\ \hline 11 & 11 \\ \hline & 1 \end{array}$$

$$\therefore\ 297\ =\ 3^3 \cdot 11$$

16.

$$\begin{array}{r|r} 3 & 1575 \\ \hline 3 & 525 \\ \hline 5 & 175 \\ \hline 5 & 35 \\ \hline 7 & 7 \\ \hline & 1 \end{array}$$

$$\therefore\ 1575\ =\ 3^2 \cdot 5^2 \cdot 7$$

17.

$$\begin{array}{r|r} 3 & 1911 \\ \hline 7 & 637 \\ \hline 7 & 91 \\ \hline 13 & 13 \\ \hline & 1 \end{array}$$

$$\therefore\ 1911\ =\ 3 \cdot 7^2 \cdot 13$$

V   18.   15 and 40

$$15 = 3 \cdot 5$$
$$40 = 2^3 \cdot 5$$
$$\text{LCM} = 2^3 \cdot 3 \cdot 5$$
$$= 8 \cdot 3 \cdot 5$$
$$= 120$$

19.   45 and 63

$$45 = 3^2 \cdot 5$$
$$63 = 3^2 \cdot 7$$
$$\text{LCM} = 3^2 \cdot 5 \cdot 7$$
$$= 9 \cdot 5 \cdot 7$$
$$= 315$$

20.   2, 9 and 12

$$2 = 2^1$$
$$9 = 3^2$$
$$12 = 2^2 \cdot 3$$
$$\text{LCM} = 2^2 \cdot 3^2$$
$$= 4 \cdot 9$$
$$= 36$$

21.   9, 14 and 21

$$9 = 3^2$$
$$14 = 2 \cdot 7$$
$$21 = 3 \cdot 7$$
$$\text{LCM} = 2 \cdot 3^2 \cdot 7$$
$$= 2 \cdot 9 \cdot 7$$
$$= 126$$

22.   28, 44 and 21

$$28 = 2^2 \cdot 7$$
$$44 = 2^2 \cdot 11$$
$$21 = 3 \cdot 7$$
$$\text{LCM} = 2^2 \cdot 3 \cdot 7 \cdot 11$$
$$= 4 \cdot 3 \cdot 7 \cdot 11$$
$$= 924$$

VI   23. $15 + 2^3 - 7$

= $15 + 8 - 7$

= $23 - 7$

= 16

24. $18 \div 3 \times 2$

= 6×2

= 12

25. $36 - 4 \div 4$

= $36 - 1$

= 35

26. $20 \div 5 + 3 \times 2$

= $4 + 6$

= 10

27. $3 \times 2^2 + 5 \times 2 - 1$

= $3 \times 4 + 5 \times 2 - 1$

= $12 + 10 - 1$

= $22 - 1$

= 21

28. $40 - 2^3 \div 4 \times 2 - 5 + 3$

= $40 - 8 \div 4 \times 2 - 5 + 3$

= $40 - 2 \times 2 - 5 + 3$

= $40 - 4 - 5 + 3$

= $36 - 5 + 3$

= $31 + 3$

= 34

VII   29. $(8 - 2) \times 4 = 6 \times 4$

$= 24$

30. $3 \times \{20 - [13 - (9 - 3)]\} = 3 \times \{20 - [13 - 6]\}$

$= 3 \times \{20 - 7\}$

$= 3 \times 13$

$= 39$

31. $12 - 2(8 - 4 \div 2) = 12 - 2(8 - 2)$

$= 12 - 2(6)$

$= 12 - 12$

$= 0$

32. $2 \times [36 \div (18 - 9) + 25 - 3^2] = 2 \times [36 \div 9 + 25 - 3^2]$

$= 2 \times [36 \div 9 + 25 - 9]$

$= 2 \times [4 + 25 - 9]$

$= 2 \times [29 - 9]$

$= 2 \times 20$

$= 40$

33. $12 \div 3 + 3 \{3 + [3 + (17 - 8) \div 3]\} = 12 \div 3 + 3 \{3 + [3 + 9 \div 3]\}$

$= 12 \div 3 + 3 \{3 + [3 + 3]\}$

$= 12 \div 3 + 3 \{3 + 6\}$

$= 12 \div 3 + 3 \{9\}$

$= 4 + 3 \{9\}$

$= 4 + 27$

$= 31$

Answers to Assignments

I   1. 8^2 3. $9^3 \cdot 21^4$

II	Base	Exponent	Value
5.	3	5	243
7.	10	2	
	2	6	100 x 64 = 6400

III   9.   5   11.   13

IV   13.

$$
\begin{array}{r|r}
2 & 54 \\
3 & 27 \\
3 & 9 \\
3 & 3 \\
& 1
\end{array}
$$

$54 = 2 \cdot 3^3$

15.

$$
\begin{array}{r|r}
3 & 1683 \\
3 & 561 \\
11 & 187 \\
17 & 17 \\
& 1
\end{array}
$$

$1683 = 3^2 \cdot 11 \cdot 17$

17.

$$
\begin{array}{r|r}
3 & 3927 \\
7 & 1309 \\
11 & 187 \\
17 & 17 \\
& 1
\end{array}
$$

$3927 = 3 \cdot 7 \cdot 11 \cdot 17$

V   19. $14 = 2 \cdot 7$

$42 = 2 \cdot 3 \cdot 7$

$LCM = 2 \cdot 3 \cdot 7$

$= 42$

21. $36 = 2^2 \cdot 3^2$

$54 = 2 \cdot 3^3$

$108 = 2^2 \cdot 3^3$

$LCM = 2^2 \cdot 3^3$

$= 4 \cdot 27$

$= 108$

VI   23.   81   VII   29.   6

25.   29   31.   5

27.   8   33.   24

Unit 3

Solutions to Drill Exercises

I   1. $\dfrac{7 \times 5}{9 \times 5} = \dfrac{35}{45}$ 2. $\dfrac{2 \times 13}{7 \times 13} = \dfrac{26}{91}$ 3. $\dfrac{11 \times 5}{32 \times 5} = \dfrac{55}{160}$

II   4. $\dfrac{6}{12} = \dfrac{6 \div 6}{12 \div 6} = \dfrac{1}{2}$ 5. $\dfrac{40}{64} = \dfrac{40 \div 8}{64 \div 8} = \dfrac{5}{8}$

6. $\dfrac{21}{72} = \dfrac{21 \div 3}{72 \div 3} = \dfrac{7}{24}$ 7. $\dfrac{184}{345} = \dfrac{184 \div 23}{345 \div 23} = \dfrac{8}{15}$

III   8. $\dfrac{2}{3}$ and $\dfrac{18}{27}$ are equal since 2 x 27 = 54 and 3 x 18 = 54.

9. $\dfrac{8}{14}$ and $\dfrac{36}{63}$ are equal since 8 x 63 = 504 and 14 x 36 = 504.

10. $\dfrac{6}{27}$ and $\dfrac{9}{41}$ are not equal since 6 x 41 = 246 and 9 x 27 = 243.

IV   11. $\dfrac{5}{9}$, $\dfrac{3}{5}$

$9 = 3^2$

$5 = 5^1$ LCD $= 3^2 \cdot 5 = 45$

12. $\dfrac{3}{8}$, $\dfrac{7}{12}$

$8 = 2^3$

$12 = 2^2 \cdot 3$ LCD $= 2^3 \cdot 3 = 24$

13. $\dfrac{2}{3}$, $\dfrac{1}{4}$, $\dfrac{5}{6}$

$3 = 3^1$

$4 = 2^2$

$6 = 2 \cdot 3$ LCD $= 3 \cdot 2^2 = 12$

14. $\dfrac{1}{3}$, $\dfrac{7}{15}$, $\dfrac{5}{18}$

$3 = 3^1$

$15 = 3 \cdot 5$

$18 = 2 \cdot 3^2$ LCD $= 2 \cdot 3^2 \cdot 5 = 90$

15. $\dfrac{4}{27}$, $\dfrac{5}{63}$, $\dfrac{1}{72}$

$27 = 3^3$

$63 = 3^2 \cdot 7$

$72 = 2^3 \cdot 3^2$ LCD $= 2^3 \cdot 3^3 \cdot 7 = 1512$

V   16. $\dfrac{3}{4}$, $\dfrac{4}{3}$

$\dfrac{3}{4} = \dfrac{3 \times 3}{4 \times 3} = \dfrac{9}{12}$

$\dfrac{4}{3} = \dfrac{4 \times 4}{3 \times 4} = \dfrac{16}{12}$ $\therefore \dfrac{3}{4}$ is the smallest fraction

17. $\dfrac{13}{20}$, $\dfrac{20}{28}$, $\dfrac{8}{14}$

$\dfrac{13}{20} = \dfrac{13 \times 7}{20 \times 7} = \dfrac{91}{140}$

$\dfrac{20}{28} = \dfrac{20 \times 5}{28 \times 5} = \dfrac{100}{140}$

$\dfrac{8}{14} = \dfrac{8 \times 10}{14 \times 10} = \dfrac{80}{140}$ $\therefore \dfrac{8}{14}$ is the smallest fraction

VI   18. $\dfrac{2}{13}$, $\dfrac{5}{32}$, $\dfrac{7}{42}$

$\dfrac{2}{13} = \dfrac{2 \times 672}{13 \times 672} = \dfrac{1344}{8736}$

$\dfrac{5}{32} = \dfrac{5 \times 273}{32 \times 273} = \dfrac{1365}{8736}$

$\dfrac{7}{42} = \dfrac{7 \times 208}{42 \times 208} = \dfrac{1456}{8736}$

$\therefore \dfrac{7}{42}$ is the largest fraction.

VI   19. $\dfrac{11}{32}$, $\dfrac{27}{80}$, $\dfrac{8}{24}$

$\dfrac{11}{32} = \dfrac{11 \times 15}{32 \times 15} = \dfrac{165}{480}$

$\dfrac{27}{80} = \dfrac{27 \times 6}{80 \times 6} = \dfrac{162}{480}$

$\dfrac{8}{24} = \dfrac{8 \times 20}{24 \times 20} = \dfrac{160}{480}$

$\therefore \dfrac{11}{32}$ is the largest fraction.

VII   20. $5 + 7 = 12$ light bulbs in total.

$\dfrac{40W \ bulbs}{total \ bulbs} = \dfrac{5}{12}$

21.   3740:220   or   17:1

22. $\dfrac{288}{16} = \dfrac{18}{1}$ $\therefore$ 18 students/section

Answers to Assignments

I   1. $\dfrac{3}{8} = \dfrac{12}{32}$

3. $\dfrac{3}{16} = \dfrac{27}{144}$

II   5. $\dfrac{36}{60} = \dfrac{3}{5}$

7. $\dfrac{360}{378} = \dfrac{20}{21}$

9. $\dfrac{5}{8}$ and $\dfrac{13}{21}$ are not equal since $5 \times 21 = 105$ and $8 \times 13 = 104$.

IV   11.   LCD = 33

13.   LCD = 24

15.   LCD = 546

V   17. $\dfrac{5}{8} = \dfrac{75}{120}$

$\dfrac{8}{12} = \dfrac{80}{120}$

$\dfrac{8}{15} = \dfrac{64}{120}$

$\therefore \dfrac{8}{12}$ is the largest fraction.

VI   19. $\dfrac{11}{24} = \dfrac{220}{480}$

$\dfrac{17}{32} = \dfrac{255}{480}$

$\dfrac{23}{40} = \dfrac{276}{480}$

$\therefore \dfrac{11}{24}$ is the smallest fraction.

VII   21.   798:38   or   21:1

Unit 4

Solutions to Drill Exercises

I   1. $4\frac{3}{7} = \frac{(4 \times 7) + 3}{7} = \frac{28 + 3}{7} = \frac{31}{7}$ 2. $1\frac{9}{11} = \frac{(1 \times 11) + 9}{11} = \frac{11 + 9}{11} = \frac{20}{11}$

3. $9\frac{2}{3} = \frac{(9 \times 3) + 2}{3} = \frac{27 + 2}{3} = \frac{29}{3}$

II   4. $\frac{48}{9} = 5\frac{3}{9} = 5\frac{1}{3}$ 5. $\frac{27}{5} = 5\frac{2}{5}$ 6. $\frac{40}{12} = 3\frac{4}{12} = 3\frac{1}{3}$

III   7. $\frac{10}{14} - \frac{3}{14} = \frac{7}{14} = \frac{1}{2}$

8. $\begin{array}{r} \frac{3}{10} \\ +\frac{5}{10} \\ \hline \frac{8}{10} = \frac{4}{5} \end{array}$

9. $\frac{2}{5} + \frac{3}{7}$

$= \frac{14}{35} + \frac{15}{35}$

$= \frac{29}{35}$

10. $\frac{7}{9} - \frac{2}{3} = \frac{7}{9} - \frac{6}{9} = \frac{1}{9}$

11. $\frac{3}{4} + \frac{1}{9} - \frac{7}{12} = \frac{27}{36} + \frac{4}{36} - \frac{21}{36}$

$= \frac{10}{36}$

$= \frac{5}{18}$

12. $\begin{array}{r} \frac{5}{24} = \frac{15}{72} \\ +\frac{2}{9} = +\frac{16}{72} \\ \hline \frac{31}{72} \end{array}$

13. $\begin{array}{r} \frac{11}{15} = \frac{66}{90} \\ -\frac{7}{18} = -\frac{35}{90} \\ \hline \frac{31}{90} \end{array}$

14. $8\frac{3}{7} + 1\frac{4}{7} = 9\frac{7}{7} = 10$

15. $12\frac{9}{10} - 7\frac{2}{5} = 12\frac{9}{10} - 7\frac{4}{10}$

$= 5\frac{5}{10}$

$= 5\frac{1}{2}$

16. $3 - 1\frac{1}{7} = 2\frac{7}{7} - 1\frac{1}{7} = 1\frac{6}{7}$ 17. $7 - \frac{1}{9} = 6\frac{9}{9} - \frac{1}{9} = 6\frac{8}{9}$

III   18. $3\frac{3}{8} = 3\frac{3}{8}$

$+\frac{3}{4} = +\frac{6}{8}$

$3\frac{9}{8} = 4\frac{1}{8}$

19. $5\frac{1}{7} = 5\frac{5}{35} = 4\frac{40}{35}$

$-2\frac{4}{5} = -2\frac{28}{35} = -2\frac{28}{35}$

$\phantom{-2\frac{4}{5} = -2\frac{28}{35} = }2\frac{12}{35}$

20. $9\frac{5}{12} = 9\frac{50}{120} = 8\frac{170}{120}$

$-7\frac{23}{40} = -7\frac{69}{120} = -7\frac{69}{120}$

$\phantom{-7\frac{23}{40} = -7\frac{69}{120} = }1\frac{101}{120}$

21. $5\frac{1}{2} - 1\frac{3}{5} + 2\frac{1}{7} = 5\frac{35}{70} - 1\frac{42}{70} + 2\frac{10}{70} = 6\frac{3}{70}$

22. $10\frac{10}{21} + 2\frac{1}{6} - 2\frac{9}{14} - 3\frac{1}{2} = \frac{220}{21} + \frac{13}{6} - \frac{37}{14} - \frac{7}{2}$

$= \frac{440}{42} + \frac{91}{42} - \frac{111}{42} - \frac{147}{42}$

$= \frac{273}{42}$

$= 6\frac{21}{42} = 6\frac{1}{2}$

IV   23. $\frac{1}{3} + \frac{2}{5} = \frac{5}{15} + \frac{6}{15} = \frac{11}{15}$

$1 - \frac{11}{15} = \frac{4}{15}$

$\therefore \frac{4}{15}$ of the alloy is copper.

24. $1\frac{1}{2} - \frac{1}{4} = 1\frac{2}{4} - \frac{1}{4} = 1\frac{1}{4}$

$\therefore$ she spent $1\frac{1}{4}$ hours studying.

25. $7\frac{3}{4} + 8\frac{1}{3} + 8\frac{1}{2} + 9\frac{1}{4} + 7\frac{2}{3}$

$= 7\frac{9}{12} + 8\frac{4}{12} + 8\frac{6}{12} + 9\frac{3}{12} + 7\frac{8}{12}$

$= 39\frac{30}{12}$

$= 39 + 2\frac{6}{12}$

$= 41\frac{1}{2}$

$\therefore$ he worked $41\frac{1}{2}$ hours that week.

26. $\frac{1}{3} + \frac{2}{5} + \frac{1}{10} = \frac{10}{30} + \frac{12}{30} + \frac{3}{30} = \frac{25}{30}$

$1 - \frac{25}{30} = \frac{5}{30} = \frac{1}{6}$

$\therefore \frac{1}{6}$ of the shares were Exon stock.

27. $\frac{3}{8} + \frac{1}{6} + \frac{1}{3} = \frac{9}{24} + \frac{4}{24} + \frac{8}{24} = \frac{21}{24}$

$1 - \frac{21}{24} = \frac{3}{24} = \frac{1}{8}$

$\therefore$ there was $\frac{1}{8}$ of a tank of gas left.

Answers to Assignments

I 1. $\dfrac{16}{5}$ 3. $\dfrac{59}{8}$

II 5. $7\dfrac{2}{3}$

III 7. $\dfrac{1}{2}$ 9. $4\dfrac{3}{5}$ 11. $1\dfrac{4}{35}$ 13. $\dfrac{2}{9}$

   15. $\dfrac{14}{45}$ 17. $7\dfrac{3}{4}$ 19. $2\dfrac{2}{35}$ 21. $7\dfrac{7}{60}$

IV 23. $\dfrac{1}{12}$ of the students.   25. $\dfrac{79}{150}$ cm.

Unit 5

Solutions to Drill Exercises

I 1. $16 \times \dfrac{2}{3} = \dfrac{32}{3} = 10\dfrac{2}{3}$

2. $\dfrac{11}{\overset{}{\underset{6}{\cancel{12}}}} \times \overset{5}{\cancel{10}} = \dfrac{55}{6} = 9\dfrac{1}{6}$

3. $\overset{4}{\cancel{32}} \times \dfrac{3}{\underset{1}{\cancel{8}}} = 12$

4. $\overset{20}{\cancel{100}} \times \dfrac{3}{\underset{1}{\cancel{5}}} = 60$

5. $\dfrac{1}{4}$ of $\dfrac{3}{7} = \dfrac{1}{4} \times \dfrac{3}{7} = \dfrac{3}{28}$

6. $\dfrac{4}{5} \times \dfrac{2}{5} = \dfrac{8}{25}$

7. $\dfrac{\overset{3}{\cancel{6}}}{\underset{2}{\cancel{14}}} \times \dfrac{\overset{3}{\cancel{21}}}{\underset{1}{\cancel{2}}} = \dfrac{9}{2} = 4\dfrac{1}{2}$

8. $\dfrac{\overset{5}{\cancel{10}}}{\underset{3}{\cancel{9}}} \times \dfrac{\overset{5}{\cancel{15}}}{\underset{4}{\cancel{8}}} \times \dfrac{25}{12} = 2\dfrac{1}{12}$

9. $\dfrac{\overset{1}{\cancel{3}}}{\underset{1}{\cancel{5}}} \times \dfrac{\overset{\overset{1}{\cancel{4}}}{\cancel{20}}}{\underset{\underset{3}{\cancel{12}}}{\cancel{36}}} \times \dfrac{1}{3}$

10. $1\dfrac{1}{2} \times 2\dfrac{2}{3} = \dfrac{\overset{1}{\cancel{3}}}{\underset{1}{\cancel{2}}} \times \dfrac{\overset{4}{\cancel{8}}}{\underset{1}{\cancel{3}}} = 4$

11. $\dfrac{2}{3} \times 3\dfrac{9}{10} = \dfrac{\overset{1}{\cancel{2}}}{\underset{1}{\cancel{3}}} \times \dfrac{\overset{13}{\cancel{39}}}{\underset{5}{\cancel{10}}} = \dfrac{13}{5}$
   $= 2\dfrac{3}{5}$

12. $4\dfrac{3}{5} \times 8 = \dfrac{23}{5} \times 8 = \dfrac{184}{5}$
    $= 36\dfrac{4}{5}$

13. $2\dfrac{1}{3} \times 2\dfrac{1}{7} \times 3\dfrac{1}{2} = \dfrac{\overset{1}{\cancel{7}}}{\underset{1}{\cancel{3}}} \times \dfrac{\overset{5}{\cancel{15}}}{\underset{1}{\cancel{7}}} \times \dfrac{7}{2}$
    $= \dfrac{35}{2}$
    $= 17\dfrac{1}{2}$

14. $4\dfrac{2}{5} \times \dfrac{2}{3} \times 1\dfrac{3}{4} = \dfrac{\overset{11}{\cancel{22}}}{5} \times \dfrac{\overset{1}{\cancel{2}}}{3} \times \dfrac{7}{\underset{\underset{1}{\cancel{2}}}{\cancel{4}}}$
    $= \dfrac{77}{15}$
    $= 5\dfrac{2}{15}$

I   15. $\dfrac{1}{\cancel{4}_1} \times \dfrac{2}{3} \times \cancel{\dfrac{2}{8}} = \dfrac{4}{3} = 1\dfrac{1}{3}$

II   16. $\left(\dfrac{1}{4}\right)^3 = \dfrac{1}{4} \cdot \dfrac{1}{4} \cdot \dfrac{1}{4}$

$= \dfrac{1}{64}$

17. $\left(\dfrac{2}{7}\right)^2 = \dfrac{2}{7} \cdot \dfrac{2}{7}$

$= \dfrac{4}{49}$

18. $\left(\dfrac{1}{3}\right)^4 = \dfrac{1}{3} \cdot \dfrac{1}{3} \cdot \dfrac{1}{3} \cdot \dfrac{1}{3}$

$= \dfrac{1}{81}$

19. $\left(\dfrac{3}{8}\right)^2 = \dfrac{3}{8} \cdot \dfrac{3}{8}$

$= \dfrac{9}{64}$

20. $\sqrt{\dfrac{1}{4}} = \sqrt{\dfrac{1^2}{2^2}} = \dfrac{1}{2}$

22. $\sqrt{\dfrac{25}{49}} = \sqrt{\dfrac{5^2}{7^2}} = \dfrac{5}{7}$

21. $\sqrt{\dfrac{36}{144}} = \sqrt{\dfrac{6^2}{12^2}} = \dfrac{6}{12} = \dfrac{1}{2}$

23. $\sqrt{\dfrac{16}{121}} = \sqrt{\dfrac{4^2}{11^2}} = \dfrac{4}{11}$

III   24. The reciprocal of $\dfrac{3}{11}$ is $\dfrac{11}{3}$.

25. $17 = \dfrac{17}{1}$, hence the reciprocal of 17 is $\dfrac{1}{17}$.

26. $2\dfrac{1}{5} = \dfrac{11}{5}$, hence the reciprocal of $2\dfrac{1}{5}$ is $\dfrac{5}{11}$.

IV   27. $\dfrac{7}{12} \div \dfrac{2}{3} = \dfrac{7}{\cancel{12}_4} \times \dfrac{\cancel{3}^1}{2} = \dfrac{7}{8}$

28. $\dfrac{1}{2} \div \dfrac{1}{4} = \dfrac{1}{\cancel{2}_1} \times \dfrac{\cancel{4}^2}{1} = 2$

29. $7 \div \dfrac{3}{5} = \dfrac{7}{1} \times \dfrac{5}{3} = \dfrac{35}{3}$

$= 11\dfrac{2}{3}$

30. $\dfrac{3}{4} \div 5 = \dfrac{3}{4} \times \dfrac{1}{5} = \dfrac{3}{20}$

31. $3\dfrac{1}{2} \div 2\dfrac{2}{3} = \dfrac{7}{2} \div \dfrac{8}{3}$

$= \dfrac{7}{2} \times \dfrac{3}{8}$

$= \dfrac{21}{16}$

$= 1\dfrac{5}{16}$

32. $5\dfrac{1}{3} \div \dfrac{3}{4} = \dfrac{16}{3} \div \dfrac{3}{4}$

$= \dfrac{16}{3} \times \dfrac{4}{3}$

$= \dfrac{64}{9}$

$= 7\dfrac{1}{9}$

V   33. $\dfrac{\frac{3}{4}}{\frac{9}{12}}$ = $\dfrac{3}{4} \div \dfrac{9}{12}$

= $\dfrac{\overset{1}{\cancel{3}}}{\underset{1}{\cancel{4}}} \times \dfrac{\overset{1}{\cancel{12}}^{3}}{\underset{1}{\cancel{9}}_{3}}$

= 1

34. $\dfrac{6}{\frac{3}{5}}$ = $6 \div \dfrac{3}{5}$

= $\overset{2}{\cancel{6}} \times \dfrac{5}{\underset{1}{\cancel{3}}}$

= 10

35. $\dfrac{\frac{1}{6}}{5}$ = $\dfrac{1}{6} \div 5$

= $\dfrac{1}{6} \times \dfrac{1}{5}$

= $\dfrac{1}{30}$

36. $\dfrac{\frac{3}{4}+\frac{1}{2}}{\frac{5}{8}}$ = $\dfrac{\frac{3}{4}+\frac{2}{4}}{\frac{5}{8}}$ = $\dfrac{\frac{5}{4}}{\frac{5}{8}}$ = $\dfrac{5}{4} \div \dfrac{5}{8}$

= $\dfrac{5}{\underset{1}{\cancel{4}}} \times \dfrac{\overset{2}{\cancel{8}}}{\cancel{5}}$

= 2

37. $\dfrac{1}{7-\frac{3}{7}}$ = $\dfrac{1}{6\frac{7}{7}-\frac{3}{7}}$ = $\dfrac{1}{6\frac{4}{7}}$ = $\dfrac{1}{\frac{46}{7}}$ = $1 \div \dfrac{46}{7}$

= $1 \times \dfrac{7}{46}$

= $\dfrac{7}{46}$

38. $\dfrac{\frac{2}{3}+\frac{1}{2}}{3\frac{1}{3}-\frac{2}{3}}$ = $\dfrac{\frac{4}{6}+\frac{3}{6}}{\frac{10}{3}-\frac{2}{3}}$ = $\dfrac{\frac{7}{6}}{\frac{8}{3}}$ = $\dfrac{7}{6} \div \dfrac{8}{3}$

= $\dfrac{7}{\underset{2}{\cancel{6}}} \times \dfrac{\overset{1}{\cancel{3}}}{8}$

= $\dfrac{7}{16}$

VI   39. $\dfrac{117}{1\frac{5}{8}}$ = $\dfrac{117}{\frac{13}{8}}$ = $117 \div \dfrac{13}{8}$ = $\overset{9}{\cancel{117}} \times \dfrac{8}{\underset{1}{\cancel{13}}}$ = 72

The average speed was 72 km/hr.

40. 20 minutes = $\dfrac{20}{60}$ = $\dfrac{1}{3}$ hr.

$24 \times 1\dfrac{1}{3}$ = $\overset{8}{\cancel{24}} \times \dfrac{4}{\underset{1}{\cancel{3}}}$ = 32

In 1 hour and 20 minutes, the machine can produce 32 parts.

41. $124\dfrac{1}{2} \div 16\dfrac{3}{5}$ = $\dfrac{249}{2} \div \dfrac{83}{5}$ = $\dfrac{\overset{3}{\cancel{249}}}{2} \times \dfrac{5}{\underset{1}{\cancel{83}}}$ = $\dfrac{15}{2}$ = $7\dfrac{1}{2}$

The worker has worked $7\dfrac{1}{2}$ hours.

42. $75 \times 1\dfrac{1}{2}$ = $75 \times \dfrac{3}{2}$ = $\dfrac{225}{2}$ = $112\dfrac{1}{2}$

The voltage sum is $112\dfrac{1}{2}$ volts.

43. $67\dfrac{1}{2} \div 2\dfrac{1}{4}$ = $\dfrac{135}{2} \div \dfrac{9}{4}$ = $\dfrac{135}{\underset{1}{\cancel{2}}} \times \dfrac{\overset{2}{\cancel{4}}}{9}$ = $\dfrac{270}{9}$ = 30

Thirty of these connections can be made.

Answers to Assignments

I   1. $116\frac{2}{3}$ 3.   6   5. $12\frac{1}{2}$ 7. $1\frac{17}{18}$

     9. $\frac{15}{64}$ 11. $66\frac{2}{3}$ 13. $\frac{7}{16}$ 15. $39\frac{3}{5}$

II   17. $\frac{1}{16}$ 19. $\frac{1}{243}$ 21. $\frac{1}{10}$ 23. $\frac{2}{5}$

III   25. $\frac{7}{2}$

IV   27. $\frac{1}{3}$ 29. $\frac{1}{18}$ 31. $\frac{35}{96}$

V   33. $\frac{9}{2}$ 35. $\frac{3}{16}$ 37. $\frac{5}{18}$

VI   39. $3\frac{1}{3}$ hP   41. $3\frac{1}{2}$ grams   43. $5\frac{1}{4}$ hP

Unit 6

Solutions to Drill Exercises

I   1. 3728.45

$$= 3(1000) + 7(100) + 2(10) + 8(1) + 4\left(\frac{1}{10}\right) + 5\left(\frac{1}{100}\right)$$

$$= 3000 + 700 + 20 + 8 + \frac{4}{10} + \frac{5}{100}$$

     2. 0.7206

$$= 7\left(\frac{1}{10}\right) + 2\left(\frac{1}{100}\right) + 0\left(\frac{1}{1000}\right) + 6\left(\frac{1}{10\,000}\right)$$

$$= \frac{7}{10} + \frac{2}{100} + \frac{6}{10\,000}$$

I   3. 2.715

$$= 2(1) + 7\left(\frac{1}{10}\right) + 1\left(\frac{1}{100}\right) + 5\left(\frac{1}{1000}\right)$$

$$= 2 + \frac{7}{10} + \frac{1}{100} + \frac{5}{1000}$$

     4. 209.08

$$= 2(100) + 0(10) + 9(1) + 0\left(\frac{1}{10}\right) + 8\left(\frac{1}{100}\right)$$

$$= 200 + 9 + \frac{8}{100}$$

II   5.   3427.03 (to the nearest ten) is 3430   6.   0.7952 (to the nearest thousandth) is 0.795

     7.   227.73 (to the nearest unit) is 228   8.   64.999 (to the nearest hundredth) is 65.00

III   9.
```
  22
 0.825
 9.7
 5.93
24
+0.0032
40.4582
```
     10.
```
  1
 0.6
 0.3
+0.7
 1.6
```
     11.
```
 3 9 9 9
40.0 1 0 1 0 1 0
- 3.0 4 2
36.9 5 8
```

III   12. $\overset{2\;\;6\;\;9}{\overset{1}{3}0\overset{1}{7}.\overset{1}{0}\overset{1}{0}15}$ 13.   4.7   14.   0.07
                    -26.49 x0.38   x0.019
                    28 0. 5 2 5   376   63
                                                     1 41   7
                                                     1.786   .00133

15.   2 16   16. $(0.2)^3$ = (0.2) (0.2) (0.2) = 0.008
         x0.214
           864   17. $(1.5)^4$ = (1.5) (1.5) (1.5) (1.5) = 5.0625
         2 16
         43 2   18. $(0.135)^2$ = (0.135) (0.135) = 0.018225
         46.224

19. 0.75 of 80 = 0.75 x 80 = 60   20. 1.95 of 4.08 = 1.95 x 4.08 = 7.956

21. 3.9 x 100 = 390   22. 38.4 x 10 = 384   23.   0.0003 x 10 000 = 3

              17.39   1.63   0.172
24.   4)69.56   25.   28)45.64   26.   13)2.236
              4   28   13
             29   17 6   93
             28   16 8   91
              1 5   84   26
              1 2   84   26
              36   0   0
              36
               0

27.   0.9)16.2.   28.   0.06)36.00.

              18   600
           9)162   6)3600
              9   3600
             72   0
             72
              0
         16.2 ÷ 0.9 = 18   36 ÷ 0.06 = 600

29. 0.065 ÷ 0.1 = 0.65   30. 7.263 ÷ 100 = 0.07263

31. 493.05 ÷ 10 000 = 0.049305   32. 0.037 ÷ 10 = 0.0037

IV
              9.045   2.49
33.   14)126.63   34.   77)191.73
              126   154
              63   37 7
              56   30 8
              70   6 93
              70   6 93
               0   0

         To the nearest tenth   To the nearest thousandth
         126.63 ÷ 14 = 9.0   191.73 ÷ 77 = 2.490

IV

35.
$$26\overline{)444.342}$$ = 17.090

```
        17.090
26)444.342
    26
    184
    182
      2 34
      2 34
        02
```

To the nearest hundredth
444.342 ÷ 26 = 17.09

36.
$$4.5\overline{)1758.0.}$$

```
        390.666
45)17580
   135
   408
   405
    30 0
    27 0
     3 00
     2 70
       300
       270
        30
```

To the nearest hundredth
1758 ÷ 4.5 = 390.67

37.
$$3.9\overline{)7.6.06}$$

```
     1.95
39)76.06
   39
   37 0
   35 1
    1 96
    1 95
       1
```

To the nearest tenth
7.606 ÷ 3.9 = 2.0

38.
$$1.7\overline{)6.5.3}$$

```
      3.8411
17)65.3
   51
   14 3
   13 6
      70
      68
      20
      17
       30
       17
        3
```

To the nearest thousandth
6.53 ÷ 1.7 = 3.841

V   39.   There will be 3 cuts, hence 3 x 0.7 = 2.1 cm of stock wasted in cutting.
14.9 + 11.2 + 5.25 + 2.1 = 33.45
50.8 − 33.45 = 17.35
Therefore, there are 17.35 cm of stock left.

40.   9.813 x 3.14 = 30.81282
To the nearest thousandth : 30.813
Hence, the circumference of the circle to the nearest thousandth is 30.813 cm.

41.   46 ÷ 0.0223 = 2062.7803
So, there are approximately 2063 sheets of metal in the stack.

42.
```
   0.662
   0.250
   0.125
 +1.200
   2.237
```
Therefore, the perimeter of the figure is 2.237 m.

43. $\dfrac{45 \text{ minutes}}{60 \text{ minutes}}$ = 0.75 hour

356.25 ÷ 3.75 = 95
So, the average speed was 95 km/hr.

Hence, driving time = 3.75 hours.

Answers to Assignments

I   1. $803.15 = 8(100) + 0(10) + 3(1) + 1\left(\dfrac{1}{10}\right) + 5\left(\dfrac{1}{100}\right)$

$= 800 + 3 + \dfrac{1}{10} + \dfrac{5}{100}$

3. $4.003 = 4(1) + 0\left(\dfrac{1}{10}\right) + 0\left(\dfrac{1}{100}\right) + 3\left(\dfrac{1}{1000}\right)$

$= 4 + \dfrac{3}{1000}$

II   5.   0.129 (to the nearest tenth) is 0.1

7.   6984.6991 (to the nearest hundredth) is 6984.70

III   9.   3.0   11.   499.996   13.   0.28

15.   0.08   17.   162.5   19.   0.003 052

21.   80   23.   0.001 953 125   25.   127

27.   1023   29.   30.888   31.   2396.4

IV   33.   3.6   35.   0.81   37.   0.743

V   39.   You need 20.7 m of baseboard.   41.   0.1416 cm   43.   0.043 cm oversize

Unit 7

Solutions to Drill Exercises

I

		Symbol	Meaning
1.	1 decimetre	1 dm	0.1 m
2.	1 megametre	1 Mm	1 000 000 m
3.	1 micrometre	1 μm	0.000 001 m
4.	1 kilometre	1 km	1000 m
5.	1 decametre	1 dam	10 m
6.	1 millimetre	1 mm	0.001 m
7.	1 hectometre	1 hm	100 m
8.	1 centimetre	1 cm	0.01 m

II   9.

$$100 \text{ cm} = 1 \text{ m}$$
$$260.8 \text{ cm} = 260.8 \text{ cm} \times \frac{1 \text{ m}}{100 \text{ cm}}$$
$$= 2.608 \text{ m}$$

10.

$$1000 \text{ m} = 1 \text{ m}$$
$$650 \text{ mm} = 650 \text{ mm} \times \frac{1 \text{ m}}{1000 \text{ mm}}$$
$$= 0.65 \text{ m}$$

11.

$$1 \text{ km} = 1000 \text{ m}$$
$$0.025 \text{ km} = 0.025 \text{ km} \times \frac{1000 \text{ m}}{1 \text{ km}}$$
$$= 25 \text{ m}$$

12.

$$1 \text{ km} = 1000 \text{ m}$$
$$2.34 \text{ km} = 2.34 \text{ km} \times \frac{1000 \text{ m}}{1 \text{ km}}$$
$$= 2340 \text{ m}$$

II   13. $\quad$ 1 dam $\quad = \quad$ 10 m

$\qquad$ 0.75 dam $\quad = \quad$ 0.75 ~~dam~~ x $\dfrac{10 \text{ m}}{1 \text{ ~~dam~~}}$

$\qquad\qquad\qquad = \quad$ 7.5 m

14. $\quad$ 1 hm $\quad = \quad$ 100 m

$\qquad$ 9.5 hm $\quad = \quad$ 9.5 ~~hm~~ x $\dfrac{100 \text{ m}}{1 \text{ ~~hm~~}}$

$\qquad\qquad\qquad = \quad$ 950 m

III   15. $\quad$ 1000 m $\quad = \quad$ 1 km

$\qquad$ 954.8 m $\quad = \quad$ 954.8 ~~m~~ x $\dfrac{1 \text{ km}}{1000 \text{ ~~m~~}}$

$\qquad\qquad\qquad = \quad$ 0.9548 km

16. $\quad$ 100 m $\quad = \quad$ 1 hm

$\qquad$ 1000 m $\quad = \quad$ 1 km

$\qquad$ 95.6 hm $\quad = \quad$ 95.6 ~~hm~~ x $\dfrac{100 \text{ m}}{1 \text{ ~~hm~~}}$

$\qquad\qquad\qquad = \quad$ 9560 m

$\qquad$ 9560 m $\quad = \quad$ 9560 ~~m~~ x $\dfrac{1 \text{ km}}{1000 \text{ ~~m~~}}$

$\qquad\qquad\qquad = \quad$ 9.560 km

OR   using 10 hm $\quad = \quad$ 1 km

$\qquad$ 95.6 hm $\quad = \quad$ 95.6 ~~hm~~ x $\dfrac{1 \text{ km}}{10 \text{ ~~hm~~}}$

$\qquad\qquad\qquad = \quad$ 9.56 km

17. $\quad$ 1000 m $\quad = \quad$ 1 km

$\qquad$ 3054 m $\quad = \quad$ 3054 ~~m~~ x $\dfrac{1 \text{ km}}{1000 \text{ ~~m~~}}$

$\qquad\qquad\qquad = \quad$ 3.054 km

18. $\quad$ 100 m $\quad = \quad$ 1 hm

$\qquad$ 1000 m $\quad = \quad$ 1 km

$\qquad$ 43.2 hm $\quad = \quad$ 43.2 ~~hm~~ x $\dfrac{100 \text{ m}}{1 \text{ ~~hm~~}}$

$\qquad\qquad\qquad = \quad$ 4320 m

$\qquad$ 4320 m $\quad = \quad$ 4320 ~~m~~ x $\dfrac{1 \text{ km}}{1000 \text{ ~~m~~}}$

$\qquad\qquad\qquad = \quad$ 4.320 km

OR   using 10 hm $\quad = \quad$ 1 km

$\qquad$ 43.2 hm $\quad = \quad$ 43.2 ~~hm~~ x $\dfrac{1 \text{ km}}{10 \text{ ~~hm~~}}$

$\qquad\qquad\qquad = \quad$ 4.32 km

19. $\quad$ 1000 m $\quad = \quad$ 1 km

$\qquad$ 10 890 m $\quad = \quad$ 10 890 ~~m~~ x $\dfrac{1 \text{ km}}{1000 \text{ ~~m~~}}$

$\qquad\qquad\qquad = \quad$ 10.890 km

20. $\quad$ 10 m $\quad = \quad$ 1 dam

$\qquad$ 1000 m $\quad = \quad$ 1 km

$\qquad$ 169.5 dam $\quad = \quad$ 169.5 ~~dam~~ x $\dfrac{10 \text{ m}}{1 \text{ ~~dam~~}}$

$\qquad\qquad\qquad = \quad$ 1695 m

$\qquad$ 1695 m $\quad = \quad$ 1695 ~~m~~ x $\dfrac{1 \text{ km}}{1000 \text{ ~~m~~}}$

$\qquad\qquad\qquad = \quad$ 1.695 km

OR   using 100 dam $\quad = \quad$ 1 km

$\qquad$ 169.5 dam $\quad = \quad$ 169.5 ~~dam~~ x $\dfrac{1 \text{ km}}{100 \text{ ~~dam~~}}$

$\qquad\qquad\qquad = \quad$ 1.695 km

IV   21. $\quad$ 1 m $\quad = \quad$ 100 cm

$\qquad$ 4.03 m $\quad = \quad$ 4.03 ~~m~~ x $\dfrac{100 \text{ cm}}{1 \text{ ~~m~~}}$

$\qquad\qquad\qquad = \quad$ 403 cm

22. $\quad$ 10 mm $\quad = \quad$ 1 cm

$\qquad$ 32 mm $\quad = \quad$ 32 ~~mm~~ x $\dfrac{1 \text{ cm}}{10 \text{ ~~mm~~}}$

$\qquad\qquad\qquad = \quad$ 3.2 cm

IV   23. $1 \text{ dm} = 10 \text{ cm}$

$0.5 \text{ dm} = 0.5 \text{ dm} \times \dfrac{10 \text{ cm}}{1 \text{ dm}}$

$= 5 \text{ cm}$

24. $10 \text{ mm} = 1 \text{ cm}$

$126 \text{ mm} = 126 \text{ mm} \times \dfrac{1 \text{ cm}}{10 \text{ mm}}$

$= 12.6 \text{ cm}$

25. $1 \text{ m} = 100 \text{ cm}$

$0.091 \text{ m} = 0.091 \text{ m} \times \dfrac{100 \text{ cm}}{1 \text{ m}}$

$= 9.1 \text{ cm}$

26. $10 \text{ mm} = 1 \text{ cm}$

$6.7 \text{ mm} = 6.7 \text{ mm} \times \dfrac{1 \text{ cm}}{10 \text{ mm}}$

$= 0.67 \text{ cm}$

V   27. $1 \text{ cm} = 10 \text{ mm}$

$0.95 \text{ cm} = 0.95 \text{ cm} \times \dfrac{10 \text{ mm}}{1 \text{ cm}}$

$= 9.5 \text{ mm}$

28. $1 \text{ m} = 1000 \text{ mm}$

$0.08 \text{ m} = 0.08 \text{ m} \times \dfrac{1000 \text{ mm}}{1 \text{ m}}$

$= 80 \text{ mm}$

29. $1 \text{ cm} = 10 \text{ mm}$

$12.8 \text{ cm} = 12.8 \text{ cm} \times \dfrac{10 \text{ mm}}{1 \text{ cm}}$

$= 128 \text{ mm}$

30. $1 \text{ m} = 1000 \text{ mm}$

$0.75 \text{ m} = 0.75 \text{ m} \times \dfrac{1000 \text{ mm}}{1 \text{ m}}$

$= 750 \text{ mm}$

31. $1 \text{ cm} = 10 \text{ mm}$

$250 \text{ cm} = 250 \text{ cm} \times \dfrac{10 \text{ mm}}{1 \text{ cm}}$

$= 2500 \text{ mm}$

32. $1 \text{ m} = 1000 \text{ mm}$

$0.018 = 0.018 \text{ m} \times \dfrac{1000 \text{ mm}}{1 \text{ m}}$

$= 18 \text{ mm}$

Answers to Assignments

I   1.   1750 m

3.   1.525 m

5.   0.085 m

7.   4.5 m

9.   25 m

II   11.   1.705 km

13.   0.6394 km

15.   0.257 km

III   17.   75 cm

19.   2.56 cm

21.   2410 cm

IV   23.   10.9 mm

25.   64 mm

27.   68 mm

Unit 8

Solutions to Drill Exercises

I

		Symbol	Meaning
1.	1 decilitre	1 dL	0.1 L
2.	1 kilogram	1 kg	1000 g
3.	1 millilitre	1 mL	0.001 L
4.	1 tonne	1 t	1000 kg
5.	1 hectolitre	1 hL	100 L
6.	1 microgram	1 µg	0.000 001 g

I   *(continued)*   Symbol   Meaning

    7. 1 milligram   1 mg   0.001 g

    8. 1 decagram   1 dag   10 g

    9. 1 centilitre   1 cL   0.01 L

II   10.   1 km = 1000 m   11.   1 hm = 100 m

           $1 km^2$ = 1 000 000 m^2 $1 hm^2$ = 10 000 m^2

    0.04 km^2 = 0.04 $\cancel{km^2}$ x $\dfrac{1\ 000\ 000\ m^2}{1\ \cancel{km^2}}$ 0.09 hm^2 = 0.09 $\cancel{hm^2}$ x $\dfrac{10\ 000\ m^2}{1\ \cancel{hm^2}}$

                  = 40 000 m^2 = 900 m^2

    12.   100 cm = 1 m   13.   1 km = 1000 m

        10 000 cm^2 = 1 m^2 $1 km^2$ = 1 000 000 m^2

        724 cm^2 = 724 $\cancel{cm^2}$ x $\dfrac{1\ m^2}{10\ 000\ \cancel{cm^2}}$ 0.0025 km^2 = 0.0025 $\cancel{km^2}$ x $\dfrac{1\ 000\ 000\ m^2}{1\ \cancel{km^2}}$

                  = 0.0724 m^2 = 2500 m^2

    14.   100 cm = 1 m   15.   1 hm = 100 m

        10 000 cm^2 = 1 m^2 $1 hm^2$ = 10 000 m^2

        1384 cm^2 = 1384 $\cancel{cm^2}$ x $\dfrac{1\ m^2}{10\ 000\ \cancel{cm^2}}$ 1.03 hm^2 = 1.03 $\cancel{hm^2}$ x $\dfrac{10\ 000\ m^2}{1\ \cancel{hm^2}}$

                  = 0.1384 m^2 = 10 300 m^2

    16.   1 m = 100 cm   17.   10 mm = 1 cm

           $1 m^2$ = 10 000 cm^2 100 mm^2 = 1 cm^2

        0.95 m^2 = 0.95 $\cancel{m^2}$ x $\dfrac{10\ 000\ cm^2}{1\ \cancel{m^2}}$ 135 mm^2 = 135 $\cancel{mm^2}$ x $\dfrac{1\ cm^2}{100\ \cancel{mm^2}}$

                  = 9500 cm^2 = 1.35 cm^2

    18.   1 m = 100 cm

           $1 m^2$ = 10 000 cm^2

        0.0042 m^2 = 0.0042 $\cancel{m^2}$ x $\dfrac{10\ 000\ cm^2}{1\ \cancel{m^2}}$

                  = 42 cm^2

III 19.   1 dm = 10 cm   20.   1 m = 100 cm

           $1 dm^3$ = 1000 cm^3 $1 m^3$ = 1 000 000 cm^3

        84 dm^3 = 84 $\cancel{dm^3}$ x $\dfrac{1000\ cm^3}{1\ \cancel{dm^3}}$ 0.0035 m^3 = 0.0035 $\cancel{m^3}$ x $\dfrac{1\ 000\ 000\ cm^3}{1\ \cancel{m^3}}$

                  = 84 000 cm^3 = 3500 cm^3

    21.   10 mm = 1 cm   22.   10 cm = 1 dm

        1000 mm^3 = 1 cm^3 1000 cm^3 = 1 dm^3

        5400 mm^3 = 5400 $\cancel{mm^3}$ x $\dfrac{1\ cm^3}{1000\ \cancel{mm^3}}$ 750 cm^3 = 750 $\cancel{cm^3}$ x $\dfrac{1\ dm^3}{1000\ \cancel{cm^3}}$

                  = 5.4 cm^3 = 0.75 dm^3

III   23.
$$100 \text{ mm} = 1 \text{ dm}$$
$$1\ 000\ 000 \text{ mm}^3 = 1 \text{ dm}^3$$
$$5000 \text{ mm}^3 = 5000 \text{ mm}^3 \times \frac{1 \text{ dm}^3}{1\ 000\ 000 \text{ mm}^3}$$
$$= 0.005 \text{ dm}^3$$

24.
$$1 \text{ m} = 10 \text{ dm}$$
$$1 \text{ m}^3 = 1000 \text{ dm}^3$$
$$0.025 \text{ m}^3 = 0.025 \text{ m}^3 \times \frac{1000 \text{ dm}^3}{1 \text{ m}^3}$$
$$= 25 \text{ dm}^3$$

IV   25.
$$1000 \text{ cm}^3 = 1 \text{ L}$$
$$5500 \text{ cm}^3 = 5500 \text{ cm}^3 \times \frac{1 \text{ L}}{1000 \text{ cm}^3}$$
$$= 5.5 \text{ L}$$

26.
$$1000 \text{ mL} = 1 \text{ L}$$
$$500 \text{ mL} = 500 \text{ mL} \times \frac{1 \text{ L}}{1000 \text{ mL}}$$
$$= 0.5 \text{ L}$$

27.
$$1 \text{ kL} = 1000 \text{ L}$$
$$710 \text{ kL} = 710 \text{ kL} \times \frac{1000 \text{ L}}{1 \text{ kL}}$$
$$= 710\ 000 \text{ L}$$

28.
$$1 \text{ hL} = 100 \text{ L}$$
$$1.5 \text{ hL} = 1.5 \text{ hL} \times \frac{100 \text{ L}}{1 \text{ hL}}$$
$$= 150 \text{ L}$$

29.
$$10 \text{ dL} = 1 \text{ L}$$
$$75 \text{ dL} = 75 \text{ dL} \times \frac{1 \text{ L}}{10 \text{ dL}}$$
$$= 7.5 \text{ L}$$

30.
$$1 \text{ dm}^3 = 1 \text{ L}$$
$$6.8 \text{ dm}^3 = 6.8 \text{ dm}^3 \times \frac{1 \text{ L}}{1 \text{ dm}^3}$$
$$= 6.8 \text{ L}$$

31.
$$1 \text{ cm}^3 = 1 \text{ mL}$$

32.
$$1 \text{ cL} = 10 \text{ mL}$$
$$5.4 \text{ cL} = 5.4 \text{ cL} \times \frac{10 \text{ mL}}{1 \text{ cL}}$$
$$= 54 \text{ mL}$$

33.
$$1 \text{ L} = 1000 \text{ mL}$$
$$0.05 \text{ L} = 0.05 \text{ L} \times \frac{1000 \text{ mL}}{1 \text{ L}}$$
$$= 50 \text{ mL}$$

34.
$$1000 \text{ μL} = 1 \text{ mL}$$
$$375 \text{ μL} = 375 \text{ μL} \times \frac{1 \text{ mL}}{1000 \text{ μL}}$$
$$= 0.375 \text{ mL}$$

35.
$$1 \text{ dm}^3 = 1 \text{ L}$$
$$1 \text{ L} = 1000 \text{ mL}$$
$$0.3 \text{ dm}^3 = 0.3 \text{ dm}^3 \times \frac{1 \text{ L}}{1 \text{ dm}^3} \times \frac{1000 \text{ mL}}{1 \text{ L}}$$
$$= 300 \text{ mL}$$

36.
$$1 \text{ cm}^3 = 1 \text{ mL}$$
$$1000 \text{ cm}^3 = 1000 \text{ cm}^3 \times \frac{1 \text{ mL}}{1 \text{ cm}^3}$$
$$= 1000 \text{ mL}$$

V   37.
$$1000 \text{ g} = 1 \text{ kg}$$
$$7250 \text{ g} = 7250 \text{ g} \times \frac{1 \text{ kg}}{1000 \text{ g}}$$
$$= 7.25 \text{ kg}$$

38.
$$1 \text{ t} = 1000 \text{ kg}$$
$$0.05 \text{ t} = 0.05 \text{ t} \times \frac{1000 \text{ kg}}{1 \text{ t}}$$
$$= 50 \text{ kg}$$

39.
$$100 \text{ cg} = 1 \text{ g}$$
$$1000 \text{ g} = 1 \text{ kg}$$
$$310 \text{ cg} = 310 \text{ cg} \times \frac{1 \text{ g}}{100 \text{ cg}} \times \frac{1 \text{ kg}}{1000 \text{ g}}$$
$$= 0.0031 \text{ kg}$$

OR using
$$100\ 000 \text{ cg} = 1 \text{ kg}$$
$$310 \text{ cg} = 310 \text{ cg} \times \frac{1 \text{ kg}}{100\ 000 \text{ cg}}$$
$$= 0.0031 \text{ kg}$$

V   40. $1 \text{ hg} = 100 \text{ g}$ OR using $10 \text{ hg} = 1 \text{ kg}$

$1000 \text{ g} = 1 \text{ kg}$ $0.9 \text{ hg} = 0.9 \text{ hg} \times \dfrac{1 \text{ kg}}{10 \text{ hg}}$

$0.9 \text{ hg} = 0.9 \text{ hg} \times \dfrac{100 \text{ g}}{1 \text{ hg}} \times \dfrac{1 \text{ kg}}{1000 \text{ g}}$ $= 0.09 \text{ kg}$

$= 0.9 \times 100 \times \dfrac{1 \text{ kg}}{1000}$

$= 0.09 \text{ kg}$

41. $1 \text{ t} = 1000 \text{ kg}$ 42. $1000 \text{ g} = 1 \text{ kg}$

$7.5 \text{ t} = 7.5 \text{ t} \times \dfrac{1000 \text{ kg}}{1 \text{ t}}$ $500 \text{ g} = 500 \text{ g} \times \dfrac{1 \text{ kg}}{1000 \text{ g}}$

$= 7500 \text{ kg}$ $= 0.5 \text{ kg}$

43. $100 \text{ cg} = 1 \text{ g}$ 44. $1 \text{ kg} = 1000 \text{ g}$

$5.5 \text{ cg} = 5.5 \text{ cg} \times \dfrac{1 \text{ g}}{100 \text{ cg}}$ $0.04 \text{ kg} = 0.04 \text{ kg} \times \dfrac{1000 \text{ g}}{1 \text{ kg}}$

$= 0.055 \text{ g}$ $= 40 \text{ g}$

45. $1000 \text{ mg} = 1 \text{ g}$ 46. $1 \text{ kg} = 1000 \text{ g}$

$2550 \text{ mg} = 2550 \text{ mg} \times \dfrac{1 \text{ g}}{1000 \text{ mg}}$ $1.8 \text{ kg} = 1.8 \text{ kg} \times \dfrac{1000 \text{ g}}{1 \text{ kg}}$

$= 2.55 \text{ g}$ $= 1800 \text{ g}$

47. $1000 \text{ mg} = 1 \text{ g}$ 48. $1 \text{ hg} = 100 \text{ g}$

$750 \text{ mg} = 750 \text{ mg} \times \dfrac{1 \text{ g}}{1000 \text{ mg}}$ $0.6 \text{ hg} = 0.6 \text{ hg} \times \dfrac{100 \text{ g}}{1 \text{ hg}}$

$= 0.75 \text{ g}$ $= 60 \text{ g}$

VI   49.   Change centimetres to metres using $100 \text{ cm} = 1 \text{ m}$.   50.   Volume $=$ length x width x height

$310 \text{ cm} = 310 \text{ cm} \times \dfrac{1 \text{ m}}{100 \text{ cm}} = 3.1 \text{ m}$ $= 25 \text{ cm} \times 10 \text{ cm} \times 5 \text{ cm}$

$= 1250 \text{ cm}^3$

$250 \text{ cm} = 250 \text{ cm} \times \dfrac{1 \text{ m}}{100 \text{ cm}} = 2.5 \text{ m}$ $= 1250 \text{ cm}^3$

Area $=$ length x width 1 cm^3 weighs 1.25 g

$=$ $3.1 \text{ m} \times 2.5 \text{ m}$ 1250 cm^3 weighs $1250 \times 1.25 \text{ g} = 1562.5 \text{ g}$

$=$ 7.75 m^2 Hence, the brick weighs 1562.5 g.

Hence, the area of the patio is 7.75 m^2.

51.   Volume $=$ length x width x height

$=$ $20 \text{ cm} \times 15 \text{ cm} \times 4 \text{ cm}$

$=$ 1200 cm^3

$1000 \text{ cm}^3 = 1 \text{ L}$

$1200 \text{ cm}^3 = 1200 \text{ cm}^3 \times \dfrac{1 \text{ L}}{1000 \text{ cm}^3}$

$=$ 1.2 L

The capacity of the box is 1.2 L.

Answers to Assignments

I 1. 100 000 m^2 3. 0.0943 m^2 5. 0.2350 m^2

   7. 24.75 cm^2 9. 340 000 cm^2

II 11. 2.574 cm^3 13. 1.2 dm^3 15. 0.0075 dm^3

III 17. 820 000 L   19. 0.025 L   21. 48.5 L

   23. 19.5 mL   25. 1300 mL   27. 790 mL

IV 29. 0.1034 kg   31. 0.454 kg   33. 4 kg

   35. 0.525 g   37. 80 g   39. 1.5 g

V 41. The area of the shed floor is 6 m^2.

Unit 9

Solutions to Drill Exercises

I 1. $\dfrac{138}{100}$ = 138%

   2. $\dfrac{\frac{1}{3}}{100}$ = $\dfrac{1}{3}$%

   3. $\dfrac{0.89}{100}$ = 0.89%

   4. $\dfrac{2}{100}$ = 2%

II 5. 75% = $75 \times \dfrac{1}{100}$

         = $\dfrac{75}{100}$

         = $\dfrac{3}{4}$

   6. 40% = $40 \times \dfrac{1}{100}$

         = $\dfrac{40}{100}$

         = $\dfrac{2}{5}$

   7. $\dfrac{3}{5}$% = $\dfrac{3}{5} \times \dfrac{1}{100}$

         = $\dfrac{3}{500}$

   8. $16\dfrac{2}{3}$% = $16\dfrac{2}{3} \times \dfrac{1}{100}$

         = $\dfrac{50}{3} \times \dfrac{1}{100}$

         = $\dfrac{1}{6}$

   9. $84\dfrac{1}{4}$% = $84\dfrac{1}{4} \times \dfrac{1}{100}$

         = $\dfrac{337}{4} \times \dfrac{1}{100}$

         = $\dfrac{337}{400}$

   10. 555% = $555 \times \dfrac{1}{100}$

         = $\dfrac{555}{100}$

         = $5\dfrac{55}{100}$

         = $5\dfrac{11}{20}$

   11. 8.4% = $8.4 \times \dfrac{1}{100}$

         = $\dfrac{8.4}{100}$

         = $\dfrac{84}{1000}$

         = $\dfrac{21}{250}$

   12. 0.8% = $0.8 \times \dfrac{1}{100}$

         = $\dfrac{0.8}{100}$

         = $\dfrac{8}{1000}$

         = $\dfrac{1}{125}$

III 13.
$$\begin{array}{r} 0.25 \\ 4\overline{)1.00} \\ \underline{8} \\ 20 \\ \underline{20} \\ 0 \end{array}$$

$$\therefore \frac{1}{4} = 0.25$$

14.
$$\begin{array}{r} 0.875 \\ 8\overline{)7.00} \\ \underline{64} \\ 60 \\ \underline{56} \\ 40 \\ \underline{40} \\ 0 \end{array}$$

$$\therefore \frac{7}{8} = 0.875$$

15.
$$\begin{array}{r} 0.5625 \\ 16\overline{)9.00} \\ \underline{8\ 0} \\ 1\ 00 \\ \underline{96} \\ 40 \\ \underline{32} \\ 80 \\ \underline{80} \\ 0 \end{array}$$

$$\therefore \frac{9}{16} = 0.5625$$

16.
$$\begin{array}{r} 0.666 \\ 3\overline{)2.000} \\ \underline{1\ 8} \\ 20 \\ \underline{18} \\ 20 \\ \underline{18} \\ 2 \end{array}$$

$$\therefore \frac{2}{3} = 0.\overline{6} \quad or \quad \frac{2}{3} = 0.667$$
rounded to three decimal places.

17.
$$\begin{array}{r} 0.2727 \\ 22\overline{)6.000} \\ \underline{4\ 4} \\ 1\ 60 \\ \underline{1\ 54} \\ 60 \\ \underline{44} \\ 160 \\ \underline{154} \\ 6 \end{array}$$

$$\therefore \frac{6}{22} = 0.\overline{27} \quad or \quad \frac{6}{22} = 0.273$$
rounded to three decimal places.

18.
$$\begin{array}{r} 0.2727 \\ 11\overline{)3.000} \\ \underline{2\ 2} \\ 80 \\ \underline{77} \\ 30 \\ \underline{22} \\ 80 \\ \underline{77} \\ 3 \end{array}$$

$$\therefore \frac{3}{11} = 0.\overline{27} \quad or \quad \frac{3}{11} = 0.273$$
rounded to three decimal places.

19. $3\frac{5}{6} = \frac{23}{6}$
$$\begin{array}{r} 3.833 \\ 6\overline{)23.000} \\ \underline{18} \\ 5\ 0 \\ \underline{4\ 8} \\ 20 \\ \underline{18} \\ 20 \\ \underline{18} \\ 2 \end{array}$$

$$\therefore 3\frac{5}{6} = 3.8\overline{3} \quad or \quad 3\frac{5}{6} = 3.833$$
rounded to three decimal places.

20. $2\frac{1}{3} = \frac{7}{3}$
$$\begin{array}{r} 2.333 \\ 3\overline{)7.000} \\ \underline{6} \\ 1\ 0 \\ \underline{9} \\ 10 \\ \underline{9} \\ 10 \\ \underline{9} \\ 1 \end{array}$$

$$\therefore 2\frac{1}{3} = 2.\overline{3} \quad or \quad 2\frac{1}{3} = 2.333$$
rounded to three decimal places.

IV   21. $0.25 = 0.25 \times \dfrac{100}{100}$

$= \dfrac{25}{100}$

$= \dfrac{1}{4}$

22. $7.6 = 7.6 \times \dfrac{10}{10}$

$= \dfrac{76}{10}$

$= 7\dfrac{6}{10}$

$= 7\dfrac{3}{5}$

23. $0.875 = 0.875 \times \dfrac{1000}{1000}$

$= \dfrac{875}{1000}$

$= \dfrac{7}{8}$

24. $0.0005 = 0.0005 \times \dfrac{10\,000}{10\,000}$

$= \dfrac{5}{10\,000}$

$= \dfrac{1}{2000}$

25. $2.32 = 2.32 \times \dfrac{100}{100}$

$= \dfrac{232}{100}$

$= 2\dfrac{32}{100}$

$= 2\dfrac{8}{25}$

26. $6.01 = 6.01 \times \dfrac{100}{100}$

$= \dfrac{601}{100}$

$= 6\dfrac{1}{100}$

V   27. $25\% = 25 \times \dfrac{1}{100}$

$= 0.25$

28. $225\% = 225 \times \dfrac{1}{100}$

$= 2.25$

29. $12\dfrac{3}{4}\% = 12\dfrac{3}{4} \times \dfrac{1}{100}$

$= 12.75 \times \dfrac{1}{100}$

$= 0.1275$

30. $\dfrac{1}{2}\% = \dfrac{1}{2} \times \dfrac{1}{100}$

$= 0.5 \times \dfrac{1}{100}$

$= 0.005$

31. $0.75\% = 0.75 \times \dfrac{1}{100}$

$= 0.0075$

32. $166\dfrac{2}{3}\% = 166\dfrac{2}{3} \times \dfrac{1}{100}$

$= 166.67 \times \dfrac{1}{100}$

$= 1.67$

33. $0.05\% = 0.05 \times \dfrac{1}{100}$

$= 0.0005$

34. $2\dfrac{1}{11}\% = 2\dfrac{1}{11} \times \dfrac{1}{100}$

$= 2.09 \times \dfrac{1}{100}$

$= 0.020\,9$

VI   35. $7.4 = 740\%$

36. $0.48 = 48\%$

37. $0.06 = 6\%$

38. $0.36345 = 36.345\%$

39. $0.007 = 0.7\%$

VII   40.

$$4\overline{)3.00}$$ quotient 0.75
$$\underline{2\ 8}$$
$$20$$
$$\underline{20}$$
$$0$$

$$\therefore \frac{3}{4} = 0.75 = 75\%$$

41. $7\frac{1}{2} = \frac{15}{2}$

$$2\overline{)15.0}$$ quotient 7.5
$$\underline{14}$$
$$1\ 0$$
$$\underline{1\ 0}$$
$$0$$

$$\therefore 7\frac{1}{2} = 7.5 = 750\%$$

42.

$$5\overline{)1.0}$$ quotient 0.2
$$\underline{1\ 0}$$
$$0$$

$$\therefore \frac{1}{5} = 0.2 = 20\%$$

43.   Solution 1:

$$2\frac{1}{6} = \frac{13}{6}$$

$$6\overline{)13.0000}$$ quotient 2.1666
$$\underline{12}$$
$$1\ 0$$
$$\underline{6}$$
$$40$$
$$\underline{36}$$
$$40$$
$$\underline{36}$$
$$40$$
$$\underline{36}$$
$$4$$

$$\therefore 2\frac{1}{6} = 2.1667$$
$$= 216.67\%$$

Solution 2:

$$2\frac{1}{6} = \frac{13}{6}$$

$$6\overline{)13.00}$$ quotient 2.16
$$\underline{12}$$
$$1\ 0$$
$$\underline{6}$$
$$40$$
$$\underline{36}$$
$$4$$

$$\therefore 2\frac{1}{6} = 2.16\frac{2}{3}$$
$$= 216\frac{2}{3}\%$$

44.   Solution 1:

$$9\overline{)5.0000}$$ quotient 0.5555
$$\underline{4\ 5}$$
$$50$$
$$\underline{45}$$
$$50$$
$$\underline{45}$$
$$50$$
$$\underline{45}$$
$$5$$

$$\therefore \frac{5}{9} = 0.555$$
$$= 55.56\%$$

Solution 2:

$$9\overline{)5.00}$$ quotient 0.55
$$\underline{4\ 5}$$
$$50$$
$$\underline{45}$$
$$5$$

$$\therefore \frac{5}{9} = 0.55\frac{5}{9}$$
$$= 55\frac{5}{9}\%$$

45.   Solution 1:

$$2\frac{8}{11} = \frac{30}{11}$$

$$11\overline{)30.0000}$$ quotient 2.7272
$$\underline{22}$$
$$8\ 0$$
$$\underline{7\ 7}$$
$$30$$
$$\underline{22}$$
$$80$$
$$\underline{77}$$
$$30$$
$$\underline{22}$$
$$8$$

$$\therefore 2\frac{8}{11} = 2.7272$$
$$= 272.73\%$$

Solution 2:

$$2\frac{8}{11} = \frac{30}{11}$$

$$11\overline{)30.00}$$ quotient 2.72
$$\underline{22}$$
$$8\ 0$$
$$\underline{7\ 7}$$
$$30$$
$$\underline{22}$$
$$8$$

$$\therefore 2\frac{8}{11} = 2.72\frac{8}{11}$$
$$= 272\frac{8}{11}\%$$

VIII 46. Change 87.55% and $\frac{7}{8}$ to decimals.

87.55% = 0.8755

$\frac{7}{8}$ = 0.875

The three numbers in decimal form are:

0.869, 0.8755, 0.875

Hence, the largest number is 0.8755 and the smallest number is 0.869.

47. Change $\frac{8}{50}$ and 0.161 to percents.

$\frac{8}{50}$ = 16%

0.161 = 16.1%

The three numbers in percent form are:

16%,   16.1%,   15.95%

Hence, 0.161 = 16.1% is the largest number and 15.95% is the smallest number.

Answers to Assignments

I   1. 9%   3. 27.5%

II   5. $\frac{9}{20}$ 7. $\frac{1}{800}$ 9. $2\frac{3}{25}$

      11. $\frac{73}{500}$

III   13. 0.375   15. 0.6875   17. $0.\overline{45}$ or 0.455   19. $7.\overline{6}$ or 7.667

IV   21. $\frac{4}{5}$ 23. $\frac{43}{1000}$ 25. $\frac{1}{1250}$

V   27. 0.5   29. 0.008   31. 0.0008   33. 0.0048

VI   35. 33%   37. 2.5%   39. 0.01%

VII   41. $583.\overline{3}$% or 583.3333%   43. $66.\overline{6}$% or 66.6667%   45. $222.\overline{72}$% or 222.7273%

VIII 47. $2\frac{1}{16}$ is the largest number   2.0595 is the smallest number

---

Unit 10

Solutions to Drill Exercises

I   1. $\frac{n}{100}$ x 32 = 12   2. $\frac{40}{100}$ x n = 63   or   0.4 x n = 63

      3. $\frac{15}{100}$ x 85 = n   or   0.15 x 85 = n

II   4. n = 7 is the solution, since 6 x 7 = 42 is true.   5. n = 10 is the solution, since 10 − 2 = 8 is true.

      6. n = 27 is the solution, since $\frac{27}{3}$ = 9 is true.

III   7.   Addition Property   8.   Multiplication Property   9.   Division Property   10.   Symmetry Property

IV   11.   If $3.6 \times n = 108$

then $\dfrac{3.6 \times n}{3.6} = \dfrac{108}{3.6}$

$n = 30$

Check: If $n = 30$, then $3.6 \times n = 3.6 \times 30 = 108$.

12.   If $2.3 + n = 21.3$

then $2.3 + n - 2.3 = 21.3 - 2.3$

$n = 19$

Check: If $n = 19$, then $2.3 + n = 2.3 + 19 = 21.3$.

13.   If $0.26 = n - 1.7$, then $n - 1.7 = 0.26$

$n - 1.7 + 1.7 = 0.26 + 1.7$

$n = 1.96$

Check: If $n = 1.96$, then $n - 1.7 = 1.96 - 1.7 = 0.26$.

14.   If $77 = \dfrac{28}{100} \times n$, then $\dfrac{28}{100} \times n = 77$

$$\dfrac{\overset{1}{\cancel{100}}}{\underset{1}{\cancel{28}}} \times \dfrac{\overset{1}{\cancel{28}}}{\underset{1}{\cancel{100}}} \times n = \dfrac{\overset{25}{\cancel{100}}}{\underset{1}{\cancel{28}}} \times \overset{11}{\cancel{77}}$$

$n = 275$

Check: If $n = 275$, then $\dfrac{28}{100} \times n = \dfrac{28}{100} \times 275 = 77$.

V   15.   Find 14% of 73.

$n = 0.14 \times 73$

$n = 10.22$

Hence, 14% of 73 is 10.22.

16.   36 is what percent of 48?

$36 = n \times 0.48$

$\dfrac{36}{0.48} = \dfrac{n \times 0.48}{0.48}$

$n = 75$

Hence, 36 is 75% of 48.

17. $5\frac{1}{4}$ % of what number is 21?

$0.0525 \times n = 21$

$\dfrac{\overset{1}{\cancel{0.0525}} \times n}{\underset{1}{\cancel{0.0525}}} = \dfrac{21}{0.0525}$

$n = 400$

Hence, $5\frac{1}{4}$ % of 400 is 21.

18. $33\frac{1}{3}$ % of 69 is what number?

$\dfrac{33\frac{1}{3}}{100} \times 69 = n$

Since $\dfrac{33\frac{1}{3}}{100} = \dfrac{\cancel{100}}{\cancel{300}} = \dfrac{1}{3}$, we have

$n = \dfrac{1}{\underset{1}{\cancel{3}}} \times \cancel{69}^{23}$

$n = 23$

Hence, $33\frac{1}{3}$ % of 69 is 23.

V   19.   What percent of 500 is 35?

$$\frac{n}{100} \times \overset{5}{\cancel{500}} = 35$$

$$n \times 5 = 35$$
$$n = 7$$

Hence, 7% of 500 is 35.

21.   Find $15\frac{3}{4}$ % of $600.

$$n = 0.1575 \times 600$$

$$n = 94.5$$

Hence, $15\frac{3}{4}$ % of $600 is $94.50

22.   2.8 is what percent of 8.4?

$$2.8 = \frac{n}{100} \times 8.4$$

$$2.8 = n \times 0.084$$

$$\frac{n \times \overset{1}{\cancel{0.084}}}{\underset{1}{\cancel{0.084}}} = \frac{2.8}{0.084}$$

$$n = 33\frac{1}{3}$$

Hence, 2.8 is $33\frac{1}{3}$ % of 8.4.

VI   24.   Round $18\frac{1}{3}$ % to 20% and $31.56 to $30.

$$0.20 \times 30 = 6$$

The exact answer is $5.79

(rounded to the nearest cent).

26.   Round $4.95 to $5 and 8.7% to 10%.

$$5 = 0.10 \times n$$

$$n = 50$$

The exact answer is $56.90

(rounded to the nearest cent).

20.   78 is $8\frac{2}{3}$ % of what number?

$$78 = \frac{8\frac{2}{3}}{100} \times n$$

Since $\frac{8\frac{2}{3}}{100} = \frac{26}{300}$, we have

$$\frac{26}{300} \times n = 78$$

$$\frac{\overset{1}{\cancel{300}}}{\underset{1}{\cancel{26}}} \times \frac{\overset{1}{\cancel{26}}}{\underset{1}{\cancel{300}}} \times n = \frac{\overset{300}{\cancel{300}}}{\underset{1}{\cancel{26}}} \times \overset{3}{\cancel{78}}$$

$$n = 900$$

Hence, 78 is $8\frac{2}{3}$ % of 900.

23.   112% of what number is 17.92?

$$\frac{112}{100} \times n = 17.92$$

$$1.12 \times n = 17.92$$

$$\frac{\overset{1}{\cancel{1.12}} \times n}{\underset{1}{\cancel{1.12}}} = \frac{17.92}{1.12}$$

$$n = 16$$

Hence, 112% of 16 is 17.92.

25.   Round $28\frac{3}{4}$ % to 30% and 85.7 to 90.

$$0.30 \times n = 90$$

$$n = 300$$

The exact answer is 298.09

(rounded to two decimal places).

27.   Round $24.80 to $25 and $103.25 to $100.

$$25 = \frac{n}{100} \times 100$$

$$n = 25$$

The exact answer is 24.02%

(rounded to two decimal places).

VI   28.   Round 17.6 to 20 and 97.4 to 100.

$$20 \quad = \quad \frac{n}{100} \times 100$$

$$n \quad = \quad 20$$

The exact answer is 18.07% (rounded to two decimal places).

VII   29.   Rewrite the problem as:

$1.80 is 12% of what number?

$$1.80 \quad = \quad 0.12 \times n$$

$$n \quad = \quad 15$$

Hence, the price of the article
before the increase was $15.

30.   Rewrite the problem as:

$420 is $16\frac{2}{3}$% of what number?

$$420 = \frac{16\frac{2}{3}}{100} \times n$$

Since $\frac{16\frac{2}{3}}{100} = \frac{50}{300} = \frac{1}{6}$, we have

$$\frac{1}{6} \times n = 420$$

$$n = 2520$$

Hence, the total sales last month were $2520.

31.   Rewrite the problem as:

$$494 \quad = \quad \frac{n}{100} \times 950$$

$$n \quad = \quad 52$$

Hence, 52% of the electorate actually voted.

32.   Rewrite the problem as:

Find 25% of 1280 items.

n = 0.25 x 1280
n = 320

Hence, the daily production was increased
by 320 items.

33.   Rewrite the problem as:

Find 5% of $125 000.

n = 0.05 x 125 000
n = 6250

Hence, he received a commission of $6250.

34.   Rewrite the problem as:

Find 20% of $52.95.

n = 0.20 x 52.95
n = 10.59

Hence, the amount of discount is $10.59.

35.   Rewrite the problem as:

Find 20% of $89.50.

n = 0.20 x 89.50
  = 17.90

Hence, the discount is $17.90;

36.   Rewrite the problem as:

40% of what number is 2400?

$$0.40 \quad x \; n \quad = \quad 2400$$

$$\frac{0.40 \; x \; n}{0.40} = \frac{2400}{0.40}$$

$$n = 6000$$

Hence, the current monthly production
is 6000 items.

37.   Rewrite the problem as:

15% of what number is 90?

$$0.15 \quad x \; n \quad = \quad 90$$

$$\frac{0.15 \; x \; n}{0.15} = \frac{90}{0.15}$$

$$n = 600$$

Hence, the regular price of the appliance is $600.

38. Rewrite the problem as:

Find 15% of $78 500

n = 0.15 x 78 500
  = 11 775

Hence, you must have a down-payment of $11 775 in order to qualify for the mortgage.

39. Rewrite the problem as:

Find 7% of 9550.

n = 0.07 x 9550
  = $668.5

Cost of car plus tax = $9550 + $668.5
                     = $10 218.50

40. Rewrite the problem as:

16% of what number is $250?

$$0.16 \ \text{x} \ n = 250$$
$$\frac{0.16 \ \text{x} \ n}{0.16} = \frac{250}{0.16}$$
$$n = 1562.50$$

Hence, the amount of his investment was $1562.50

41. Rewrite the problem as:

63% of what number is 756?

$$0.63 \ \text{x} \ n = 756$$
$$\frac{0.63 \ \text{x} \ n}{0.63} = \frac{756}{0.63}$$
$$n = 1200$$

Hence, 1200 grams of solder can be made.

42. Rewrite the problem as:

Find 8% of $10.50.

n = 0.08 x 10.50

  = 0.84 .

Hence, $10.50 + $0.84 = $11.34 is his new hourly rate.

Answers to Assignments

I   1. $\frac{36}{100}$ x 92 = n

3. $\frac{20}{100}$ x n = 73

II   5. n = 7

III   7. Subtraction Property

9. Symmetry Property

IV   11. 6.71

13. 68.6

V   15. 14%

17. $66\frac{2}{3}$

19. 500

21. $466.67 (rounded to the nearest cent)

23. 50

VI   25. $4.58 (rounded to the nearest cent)

27. 264.79 (rounded to two decimal places)

VII   29. $8.20 decrease

31. $16\frac{2}{3}$ % alcohol

33. $9500

35. 540 km

37. 6% decrease

Unit 11

Solutions to Drill Exercises

I   1. $0 > -7$ 2. $-4 < +9$ 3. $0 < +11$

     4. $-4 > -10$ 5. $+35 > +23$ 6. $-13 < +5$

II   7. T   8. T   9. F

     10. T   11. F   12. T

III 13. $-(-8) = 8$ 14. $-(+7) = -7$ 15. $-[-(+13)] = -[-13]$
                                                                 $= 13$

     16. $-[-(-6)] = -[+6]$ 17. $-\{-[-(-17)]\} = -\{-[+17]\}$
                  $= -6$ $= -\{-17\}$
                                                      $= 17$

Answers to Assignments

I   1. $+6 > -1$ 3. $+15 < +24$ 5. $-7 < -1$

II   7. False   9. True   11. False

III 13. -2 15. -11 17. $+1$

Unit 12

Solutions to Drill Exercises

I   1. $(-18) + (-7) = -(18 + 7)$ 2. $(+35) + (+11) = +(35 + 11)$
                        $= -(25)$ $= +(46)$
                        $= -25$ $= 46$

     3. $0 + (+17) = 17$ 4. $(+8) + (-3) = +(8 - 3)$
                                                         $= +(5)$
                                                         $= 5$

     5. $(-10) + (+1) = -(10 - 1)$ 6. $(-15) + 0 = -15$
                        $= -(9)$
                        $= -9$

–10)

10)

(+14)

4)

(+9)

9)

(+9) – (–4)

(+9) + (+4)

III   20. $(+2) + (-19) - (-20) - (+13) - (-12)$

= $(+2) + (-19) + (+20) + (-13) + (+12)$

= $2 - 19 + 20 - 13 + 12$

= $-17 + 20 - 13 + 12$

= $-17 + 7 + 12$

= $-17 + 19$

= 2

21. $6 - 2 + 3 - 10 + 3$

= $4 + 3 - 10 + 3$

= $7 - 10 + 3$

= $-3 + 3$

= 0

22. $11 - 14 + 6 - 9 + 2$

= $19 - 23$

= -4

23. $-1 + 3 - 8 + 4 - 5$

= $2 - 8 + 4 - 5$

= $-6 + 4 - 5$

= $-2 - 5$

= -7

24. $16 - 5 + 19 + 1 - 7 - 8$

= $11 + 19 + 1 - 7 - 8$

= $30 + 1 - 7 - 8$

= $31 - 7 - 8$

= $24 - 8$

= 16

25. $-10 + 2 - 7 + 4 - 15 + 11$

= $-8 - 7 + 4 - 15 + 11$

= $-15 + 4 - 15 + 11$

= $-11 - 15 + 11$

= $-26 + 11$

= -15

26. $3 - 8 + 1 - 12 + 11 + 9$

= $-5 + 1 - 12 + 11 + 9$

= $-4 - 12 + 11 + 9$

= $-16 + 11 + 9$

= $-5 + 9$

= 4

27. $12 + 6 - 15 + 9 - 12 - 6$

= $18 - 15 + 9 - 12 - 6$

= $3 + 9 - 12 - 6$

= $12 - 12 - 6$

= -6

28. $-5 - 3 - 12 - 7 - 1 - 10$

= $-8 - 12 - 7 - 1 - 10$

= $-20 - 7 - 1 - 10$

= $-27 - 1 - 10$

= -38

29. $8 - 10 - 7 + 9 - 15 + 30$

= $-2 - 7 + 9 - 15 + 30$

= $-9 + 9 - 15 + 30$

= $-15 + 30$

= 15

30. $-1 - 6 + 8 + 11 + 3 - 9 - 6 + 4$

= $-7 + 8 + 11 + 3 - 9 - 6 + 4$

= $1 + 11 + 3 - 9 - 6 + 4$

= $12 + 3 - 9 - 6 + 4$

= $15 - 9 - 6 + 4$

= $6 - 6 + 4$

= 4

Answers to Assignments

I   1. +31   3. +5   5. −17   7. +8   9. −2

II   11. −18   13. −4   15. −7

III   17. +4   19. −1   21. 5   23. 21   25. −39

      27. −65   29. −4

Unit 13

Solutions to Drill Exercises

I   1. $(+6)(−9) = −54$ 2. $(−13)(0) = 0$ 3. $(−5)(−7) = 35$

     4. $(−8)(+4) = −32$ 5. $(+7)(+6) = 42$ 6. $(+33)(−1) = −33$

     7. $(−4)(+1) = −4$

II   8. $(−7)(−2)(−3) = (14)(−3)$ 9. $(−4)(+6)(−2)(+3) = (−24)(−2)(3)$
                       $= −42$ $= (48)(3)$
                                                                          $= 144$

     10. $(+4)(−1)(0)(+6) = 0$ 11. $(−1)(+2)(−1)(−3)(+4)(−2)$
                                                      $= (−2)(+3)(−8)$
                                                      $= (−6)(−8)$
                                                      $= 48$

III   12. $(−2)^4 = (−2)(−2)(−2)(−2)$ 13. $(−4)^4 = (−4)(−4)(−4)(−4)$
                   $= +16$ $= 256$

     14. $−3^2 = −(9)$ 15. $(−1)^{251} = −1$ 16. $−5^2 = −(25)$
             $= −9$ $= −25$

IV   17. $(+20) ÷ (−4) = −5$ 18. $\frac{+17}{−1} = −17$ 19. $\frac{−100}{−10} = 10$

     20. $\frac{−120}{0}$ is undefined   21. $(−60) ÷ (−5) = 12$

V   22. $(−22)[(−2)^4 ÷ 4]$ 23. $2\{4[5^2 − 3^2] + (−4)^3\}$
          $= (−22)[16 ÷ 4]$ $= 2\{4[25 − 9] + (−64)\}$
          $= (−22)(4)$ $= 2\{4(16) − 64\}$
          $= −88$ $= 2\{64 − 64\}$
                                                  $= 2(0)$
                                                  $= 0$

V   24. $[(-5) + 2^2 (-3) - 10^2 + 5 + (-3)] \div (2^3 - 4^2)$

        = $[-5 + 4(-3) - 100 + 5 - 3] \div (8 - 16)$

        = $[-5 - 12 - 20 - 3] \div (-8)$

        = $(-40) \div (-8)$

        =   5

     25. $[-4^2 \div 2 + (2) (6) - (-6)] + (-1) \cdot (2)$

        = $[-16 \div 2 + (2) (6) - (-6)] + (-1) \cdot (2)$

        = $[-8 + (2) (6) - (-6)] + (-1) \cdot (2)$

        = $[-8 + 12 - (-6)] + (-1) \cdot (2)$

        = $[4 + 6] + (-1) \cdot (2)$

        = $10 + (-1) \cdot (2)$

        = $(-10) \cdot (2)$

        = -20

Answers to Assignments

I   1. +21   3. −63   5. −16   7. −60

II   9. −54   11. −600

III   13. +1   15. +16

IV   17. +7   19. −9   21. +4

V   23. 102   25. 7

Unit 14

Solutions to Drill Exercises

I   1. $2z + 6$ = $2(-2) + 6$ 2. $-x^2yz$ = $-(4)^2 (6) (-2)$

            = $-4 + 6$ =   192

            =   2

     3. $x - 4z$ = $4 - 4(-2)$ 4. $5xz^2 + 2yz$ = $5(4) (-2)^2 + 2 (6) (-2)$

            = $4 + 8$ = $80 - 24$

            =   12   =   56

     5. $y(xy + 2z) - 3$ = $6[(4) (6) + 2(-2)] - 3$

                = $6[24 - 4] - 3$

                = $6[20] - 3$

                = $120 - 3$

                =   117

II   6. $13^0 = 1$ 7. $a^7 \cdot a^{14} = a^{21}$ 8. $(6y)^0 = 1$,

9. $(2b)^5 = 2^5 b^5$ 10. $y^{13} \div y^9 = y^4$ 11. $(x+y)^2 (x+y)^3 = (x+y)^5$

  $= 32 b^5$

12. $(2ab)^3 (2ab)^2 = (2ab)^5$ 13. $(3x + 5)^3 \div (3x + 5)^2 = 3x + 5$

  $= 2^5 a^5 b^5$

  $= 32 a^5 b^5$

14. $\dfrac{10^{11}}{10^9} = 10^2$ 15. $(x - 2)(x - 2)^6 = (x - 2)^7$

  $= 100$

16. $\left(\dfrac{x^3}{a^6}\right)^5 = \dfrac{x^{15}}{a^{30}}$ 17. $\left(\dfrac{-1}{x}\right)^8 = \dfrac{1}{x^8}$

18. $\left(\dfrac{3m^2 n}{2a^3 t^3}\right)^3 = \dfrac{3^3 m^6 n^3}{2^3 s^9 t^9}$ 19. $\dfrac{11^4 \cdot 11^{12}}{11^{14}} = \dfrac{11^{16}}{11^{14}}$

  $= \dfrac{27 m^6 n^3}{8 s^9 t^9}$ $= 11^2$

  $= 121$

20. $(xy)^{15} \div (xy)^3 = (xy)^{12}$ 21. $23 y^0 = 23(1)$

  $= x^{12} y^{12}$ $= 23$

22. $(y - 3)^5 \div (y - 3)^2 = (y - 3)^3$ 23. $(2a^2 b)^4 (3ab)^2 = (2^4 a^8 b^4)(3^2 a^2 b^2)$

  $= 16(9) a^{10} b^6$

24. $x^7 (3x^2 y^2 + 2z^2)^0 = x^7 \cdot 1$ $= 144 a^{10} b^6$

  $= x^7$

III   25. $4^{-2} = \dfrac{1}{4^2} = \dfrac{1}{16}$ 26. $\left(\dfrac{1}{3^3}\right)^{-2} = \dfrac{1^{-2}}{3^{-6}} = \dfrac{3^6}{1^2} = 729$

27. $\left(\dfrac{4}{5}\right)^{-3} = \dfrac{4^{-3}}{5^{-3}} = \dfrac{5^3}{4^3} = \dfrac{125}{64}$ or $1\dfrac{61}{64}$

IV   28. $a^{-4} b^6 = \dfrac{b^6}{a^4}$ 29. $\dfrac{2^3 x^3}{2^5 y^{-6}} = \dfrac{x^3 y^6}{2^2} = \dfrac{x^3 y^6}{4}$

30. $3m^{-7} n^{-6} = \dfrac{3}{m^7 n^6}$ 31. $\dfrac{p^{-3}}{q^2} = \dfrac{1}{p^3} \cdot \dfrac{1}{q^2} = \dfrac{1}{p^3 q^2}$

32. $2^3 m^{-3} n^0 s^{-4}$ 33. $5 x^0 y^{-7} z^2 = 5 \cdot 1 \cdot \dfrac{1}{y^7} \cdot z^2$

  $= 8 \cdot \dfrac{1}{m^3} \cdot 1 \cdot \dfrac{1}{s^4}$ $= \dfrac{5 z^2}{y^7}$

  $= \dfrac{8}{m^3 s^4}$

IV   34. $\dfrac{10^{-5} \cdot 10^{11}}{10^4} = \dfrac{10^6}{10^4} = 10^2 = 100$

35. $\dfrac{10^{-12} \cdot 10^4}{10^{-9} \cdot 10^3} = \dfrac{10^{-8}}{10^{-6}} = 10^{-2} = \dfrac{1}{10^2} = \dfrac{1}{100}$

36. $9x^{-7} = \dfrac{9}{x^7}$

37. $\left(\dfrac{2a^2b^0}{7c^3}\right) = \dfrac{2^2a^41^2}{7^2c^6} = \dfrac{4a^4}{49c^6}$

38. $5(2x)^{-3} = (5)(2^{-3})(x^{-3})$

$= \dfrac{5}{2^3x^3} = \dfrac{5}{8x^3}$

39. $\left(\dfrac{2x^{-1}y^3}{-5z^{-2}}\right)^{-2} = \dfrac{2^{-2}x^2y^{-6}}{(-5)^{-2}z^4}$

$= \dfrac{(-5)^2x^2}{2^2y^6z^4} = \dfrac{25x^2}{4y^6z^4}$

40. $6(x+y)^{-3} = \dfrac{6}{(x+y)^3}$

41. $\dfrac{a^{-2}}{2} = \dfrac{1}{2a^2}$

42. $(a^2 - b)^{-1} = \dfrac{1}{a^2 - b}$

43. $\left(\dfrac{-2m^2nr^{-1}}{s^{-3}t^{-1}}\right)^{-4} = \dfrac{(-2)^{-4}m^{-8}n^{-4}r^4}{s^{12}t^4}$

$= \dfrac{r^4}{(-2)^4m^8n^4s^{12}t^4}$

$= \dfrac{r^4}{16m^8n^4s^{12}t^4}$

44. $x^0y^{-2}(3z^2)^3 = \dfrac{1 \cdot 3^3 \cdot z^6}{y^2}$

$= \dfrac{27z^6}{y^2}$

45. $(m+n)^0(m^2)^{-3} = 1 \cdot m^{-6}$

$= \dfrac{1}{m^6}$

Answers to Assignments

1. +2	3. +1	5. −32	7. 1	9. x^5
11. $(a+4)^4$	13. $\dfrac{y^9}{64x^{12}}$	15. 1000	17. $72x^8y^5$	19. 5c
21. 169	23. 32	25. 1000	27. $\dfrac{1}{x^3y^2}$	29. x^7y^2
31. $\dfrac{9y^2}{z^3}$	33. $\dfrac{8}{m^4}$	35. $\dfrac{16b^{12}}{a^4}$	37. $\dfrac{m^5n^5t^5}{32q^5}$	39. $\dfrac{3}{a^2 + y}$

Unit 15

Solutions to Drill Exercises

1.   *Note:* Do not immediately substitute x = 5, y = 3, z = −1 into the expression. Simplify first using the properties of exponents. The simplified expression must have only positive exponents.

$$3(xy^{-2}z^{-3})^2 = 3x^2y^{-4}z^{-6} = \dfrac{3x^2}{y^4z^6}$$

Substitute   x = 5, y = 3, z = (−1)

$$\dfrac{3x^2}{y^4z^6} = \dfrac{3(5)^2}{(3)^4(-1)^6} = \dfrac{3(25)}{(81)(1)} = \dfrac{75}{81} = \dfrac{25}{27}$$

2. $\dfrac{4m^{-5}}{7n^{-2}} = \dfrac{4n^2}{7m^5}$

Substitute $m = 2$, $n = 3$

$\dfrac{4n^2}{7m^5} = \dfrac{4(3)^2}{7(2)^5} = \dfrac{4(9)}{7(32)} = \dfrac{9}{56}$

3. $\dfrac{x^9 y^{-3}}{x^3(x^4 y^{-2})^0} = \dfrac{x^9}{x^3 y^3} = \dfrac{x^6}{y^3}$

Substitute $x = 2$, $y = 5$

$\dfrac{x^6}{y^3} = \dfrac{2^6}{5^3} = \dfrac{64}{125}$

4. $2x(x^{-3}y^2)^{-1}(3x^{-4}y^3z^{-1})^0 = 2x(x^3 y^{-2}) = \dfrac{2x^4}{y^2}$

Substitute $x = 3$, $y = (-5)$, $z = 6$

$\dfrac{2x^4}{y^2} = \dfrac{2(3)^4}{(-5)^2} = \dfrac{2(81)}{25} = \dfrac{162}{25}$ or $6\dfrac{12}{25}$

5. $\left(\dfrac{x^{-2}y}{y^{-3}x^{-5}}\right)^{-2} = \dfrac{x^4 y^{-2}}{y^6 x^{10}} = \dfrac{1}{y^8 x^6}$

Substitute $x = (-1)$, $y = (-2)$

$\dfrac{1}{y^8 x^6} = \dfrac{1}{(-2)^8(-1)^6} = \dfrac{1}{256}$

6. $\left(\dfrac{3r^2 t^{-1}}{r^{-3}t^4}\right)\left(\dfrac{4r^4}{9t^7}\right)^{-1} = \left(\dfrac{3r^5}{t^5}\right)\left(\dfrac{9t^7}{4r^4}\right) = \dfrac{27rt^2}{4}$

Substitute $r = (-7)$, $t = 4$

$\dfrac{27rt^2}{4} = \dfrac{27(-7)(4)^2}{4} = \dfrac{27(-7)(16)}{4} = -756$

7. $9(x^2 y)^{-1}(x^2 y^3) = \dfrac{9x^2 y^3}{x^2 y} = 9y^2$

Substitute $y = 10$, $9y^2 = 9(10)^2 = 900$

8. $(-5m^2 t^3)(m^{-2}t^{-3}) = -5m^0 t^0 = -5$

The answer is −5 regardless of what the values of m and t are.

9. $\dfrac{(-10x^{16}y^{25})^0(x^{-4}y^{-2})^{-1}}{xyz} = \dfrac{1(x^4 y^2)}{xyz} = \dfrac{x^3 y}{z}$

Substitute $x = -3$, $y = 5$, $z = -10$.

$\dfrac{x^3 y}{z} = \dfrac{(-3)^3(5)}{-10} = \dfrac{(-27)(5)}{-10} = \dfrac{27}{2} = 13.5$

10. $(a^2 b^2 c^{-10})\left(\dfrac{-3a^{-1}b^2}{2c^{-3}}\right)^{-4} = (a^2 b^2 c^{-10})\dfrac{(-3)^{-4}a^4 b^{-8}}{2^{-4}c^{12}} = \dfrac{2^4 a^6}{(-3)^4 b^6 c^2} = \dfrac{16a^6}{81b^6 c^2}$

Substitute $a = 2$, $b = -1$, $c = 7$.

$= \dfrac{16a^6}{81b^6 c^2} = \dfrac{16(2)^6}{81(-1)^6(7)^2} = \dfrac{1024}{3969} = 0.258$

11. $\dfrac{3.74}{10^5}$ may be written as 3.74×10^{-5}

12. $\dfrac{KmM}{r^2}$ may be written as $KmMr^{-2}$

13. $\dfrac{V}{Re^{RC}}$ may be written as $VR^{-1}e^{-RC}$

14. $4\,230.21 = 423.021 \times 10^1$

$= 42.3021 \times 10^2$

$= 4.23021 \times 10^3$

15. $0.002\,01 = 0.002\,01 \times 10^0$

$= 0.201 \times 10^{-2}$

$= 2.01 \times 10^{-3}$

16. $3.75 = 3.75 \times 10^0$

17. $0.000\,04 = 4 \times 10^{-5}$

18. $0.31 = 3.1 \times 10^{-1}$

19. $0.063\,4 = 6.34 \times 10^{-2}$

20. $7\,053\,210 = 7.053\,21 \times 10^6$

21. $615.27 = 6.152\,7 \times 10^2$

22. $2.03 \times 10^5 = 203\ 000$

23. $4 \times 10^6 = 4\ 000\ 000$

24. $3 \times 10^{-2} = 0.03$

25. $7.102 \times 10^{-3} = 0.007\ 102$

26. $5.2 \times 10^7 = 52\ 000\ 000$

27. $9 \times 10^{-5} = 0.000\ 09$

28. $5\ 300 \times 800\ 000$

$$= 5.3 \times 10^3 \times 8 \times 10^5$$
$$= 42.4 \times 10^8$$
$$= 4.24 \times 10^9 \quad \text{(in scientific notation)}$$
$$= 4\ 240\ 000\ 000$$

29. $70\ 000 \times 0.000\ 000\ 003$

$$= 7 \times 10^4 \times 3 \times 10^{-9}$$
$$= 21 \times 10^{-5}$$
$$= 2.1 \times 10^{-4} \quad \text{(in scientific notation)}$$
$$= 0.000\ 21$$

30. $\dfrac{0.000\ 93}{0.000\ 000\ 031} = \dfrac{9.3 \times 10^{-4}}{3.1 \times 10^{-8}}$

$$= 3.0 \times 10^4 \quad \text{(in scientific notation)}$$
$$= 30\ 000$$

31. $\dfrac{0.003 \times 70\ 000}{5\ 600 \times 0.002\ 7} = \dfrac{3 \times 10^{-3} \times 7 \times 10^4}{5.6 \times 10^3 \times 2.7 \times 10^{-3}}$

$$= \dfrac{21 \times 10^1}{15.12 \times 10^0}$$
$$= 1.388\ 888\ 9 \times 10^1 \quad \text{(in scientific notation)}$$
$$= 13.888\ 889$$

32. $\dfrac{0.000\ 008\ 1 \times 600}{0.000\ 09 \times 1\ 200\ 000} = \dfrac{8.1 \times 10^{-6} \times 6 \times 10^2}{9 \times 10^{-5} \times 1.2 \times 10^6}$

$$= \dfrac{48.6 \times 10^{-4}}{10.8 \times 10^1}$$
$$= 4.5 \times 10^{-5} \quad \text{(in scientific notation)}$$
$$= 0.000\ 045$$

33. $0.023\ 641 \times 50\ 800$

$$= 2.364\ 1 \times 10^{-2} \times 5.08 \times 10^4$$
$$= 12.009\ 628 \times 10^2$$
$$= 1.200\ 962\ 8 \times 10^3 \quad \text{(in scientific notation)}$$
$$= 1\ 200.962\ 8$$

34. $2\ 314\ 000 \div 711\ 200$

$$= (2.314 \times 10^6) \div (7.112 \times 10^5)$$
$$= 0.325\ 365\ 58 \times 10^1$$
$$= 3.253\ 655\ 8 \times 10^0 \quad \text{(in scientific notation)}$$
$$= 3.253\ 655\ 8$$

35. $\dfrac{1\ 500 \times 2.54}{645.16 \times 38\ 100\ 000}$ = $\dfrac{1.5 \times 10^3 \times 2.54 \times 10^0}{6.451\ 6 \times 10^2 \times 3.81 \times 10^7}$

   = $\dfrac{3.81 \times 10^3}{24.580\ 596 \times 10^9}$

   = $.155\ 000\ 31 \times 10^{-6}$

   = $1.550\ 003\ 1 \times 10^{-7}$ (in scientific notation)

   = $0.000\ 000\ 155\ 000\ 31$

36. $\dfrac{0.005 \times 820}{128\ 000 \times 0.000\ 125}$ = $\dfrac{5 \times 10^{-3} \times 8.2 \times 10^2}{1.28 \times 10^5 \times 1.25 \times 10^{-4}}$

   = $\dfrac{41 \times 10^{-1}}{1.6 \times 10^1}$

   = 25.625×10^{-2}

   = 2.5625×10^{-1} (in scientific notation)

   = 0.25625

37. $6\ 380\ 000$ m   38. 2.998×10^8 m/s   39. 1.93×10^4 kg/m^3

40. $0.000\ 009\ 869$ atm   41. 4.7×10^{-5} cm   42. $35\ 000\ \Omega$

43. $0.000\ 013$ w/m^2 44. 9.461×10^{13} km

Answers to Assignments

I   1. $\dfrac{21}{25}$ 3. $\dfrac{1}{36}$ 5. $\dfrac{2025}{4}$ or $506\dfrac{1}{4}$

II   7. $2^{-1}\pi^{-1}d^{-1}uI$

III 9. $0.027\ 86 \times 10^{-2}$

       $0.002\ 786 \times 10^{-1}$

       2.786×10^{-4}

IV   11. $2.150\ 4 \times 10^8$ 13. $7.504\ 2 \times 10^1$ 15. 5.76×10^{-3}

V   17. 115 19. 0.01 21. $80\ 000$

VI   23. 1.0×10^{10} 25. 3.0×10^{-5} 27. 3.75×10^0

VII 29. $4.130\ 078 \times 10^2$ 31. $2.228\ 106 \times 10^3$

Unit 16

Solutions to Drill Exercises

I   1. $2a^2 - 2b^2 - 18ab + 12b^2 - 2ab + 6a^2$ 2. $x^3 + 3x^2 - 4x^2 - 12xy - 2x^3 + 9xy$

      = $8a^2 - 20ab + 10b^2$ = $-x^3 - x^2 - 3xy$

I   3. $3x^2yz + 2xz - 5x^2yz - x - 3xz$

= $-2x^2yz - xz - x$

4. $m + 3m^2 - 4p - 5p = m + 3m^2 - 9p$

II   5. $-(r + 3a) = -r - 3a$

6. $(a^3b^3 + 3ab) + (-2b + a) = a^3b^3 + 3ab - 2b + a$

7. $4x^2 + (2x - 7) = 4x^2 + 2x - 7$

8. $m^2 - (3m - 2n) = m^2 - 3m + 2n$

9. $(x - 3xy) + (-x^2 + 7z - y) = x - 3xy - x^2 + 7z - y$

10. $p^3 - q^3 - (-p^2q^2 - pq + 2p - 7)$

= $p^3 - q^3 + p^2q^2 + pq - 2p + 7$

III   11. $(a^3 + 2a^2) - (a^2 + 3a - 2)$

= $a^3 + 2a^2 - a^2 - 3a + 2$

= $a^3 + a^2 - 3a + 2$

12. $-(-2x^2 - 3x + 2) - (x^2 + 2x) + (4x^2 + x - 3)$

= $2x^2 + 3x - 2 - x^2 - 2x + 4x^2 + x - 3$

= $5x^2 + 2x - 5$

13. $3mn - [-7n + (4mn - 2)]$

= $3mn - [-7n + 4mn - 2]$

= $3mn + 7n - 4mn + 2$

= $-mn + 7n + 2$

14. $-6x - [2x^2y - (-x^2 + y - 3)]$

= $-6x - [2x^2y + x^2 - y + 3]$

= $-6x - 2x^2y - x^2 + y - 3$

= $-2x^2y - x^2 - 6x + y - 3$

15. $p^2 - \{4p - [-2p - (3p^2 - 4)] + 5p^2 - 8\}$

= $p^2 - \{4p - [-2p - 3p^2 + 4] + 5p^2 - 8\}$

= $p^2 - \{4p + 2p + 3p^2 - 4 + 5p^2 - 8\}$

= $p^2 - 4p - 2p - 3p^2 + 4 - 5p^2 + 8$

= $-7p^2 - 6p + 12$

IV   16. $2x^2 - 3x$
                       $x - 5$
      $\underline{4x^2 + 5x + 6}$
      $6x^2 + 3x + 1$

17. $2y^2 + 9y - 5$
Subtract: $\underline{5y^2 - 6y - 7}$

$2y^2 + 9y - 5$
Add: $\underline{-5y^2 + 6y + 7}$
$-3y^2 + 15y + 2$

18. $a^3 + 6a^2 + 3$
Subtract: $\underline{-a^3 + 4a - 1}$

$a^3 + 6a^2 + 3$
Add: $\underline{a^3 - 4a + 1}$
$2a^3 + 6a^2 - 4a + 4$

19. $a^4 - 5a^3 - 3a$
                $2a^2 - 4a + 1$
      $\underline{a^4 - a^2 - 2}$
      $2a^4 - 5a^3 + a^2 - 7a - 1$

IV   20. $3x^4 \qquad -5x^2 - 2x + 3$ $3x^4 \qquad -5x^2 - 2x + 3$

Subtract: $\underline{\qquad x^3 - 2x^2 \qquad\; - 3}$ *Add:* $\underline{\qquad - x^3 + 2x^2 \qquad\; + 3}$

$\qquad\qquad\qquad\qquad\qquad\qquad\qquad\qquad\qquad\qquad\qquad\quad 3x^4 - x^3 - 3x^2 - 2x + 6$

Answers to Assignments

I   1. $7m^2 - 3mn - n^2$ 3. $x^2y^3 - xy^2 - xy^3 + x^3 - x^2y^2 + x^2y$

II   5. $x^3y^3 + xy - y + 3x$ 7. $p^3 + p^2 - pq + q + 4$ 9. $x^3 - 2xy - 3x^2 + y^2 - 4x$

III   11. $x^2 + 2x - 4$ 13. $-2a^2 + 10ab - 6$ 15. $y^3 + 9y^2 - y - 13$

IV   17. $4x^2 + 3x + 1$

Unit 17

Solutions to Drill Exercises

I   1.   Identity. True for every value of x.   2.   Conditional. True only for x = 3.

     3.   Conditional. True only for y = 6.   4.   Identity. True for every value of x.

     5.   Identity. True for every value of x.

     6.
$$4y = 28$$
$$\frac{4y}{4} = \frac{28}{4}$$
$$y = 7$$

     7.
$$16 = 2x + 4$$
$$-2x = 4 - 16$$
$$-2x = -12$$
$$\frac{-2x}{-2} = \frac{-12}{-2}$$
$$x = 6$$

     8.
$$5w + 2 = w + 10$$
$$5w - w = 10 - 2$$
$$4w = 8$$
$$\frac{4w}{4} = \frac{8}{4}$$
$$w = 2$$

     9.
$$8x - 9 = 4x - 7$$
$$8x - 4x = -7 + 9$$
$$4x = 2$$
$$\frac{4x}{4} = \frac{2}{4}$$
$$x = \frac{2}{4}$$
$$x = \frac{1}{2}$$

I   10. $7w - 8 - 4w = 20 - w + 4$

$3w - 8 = 24 - w$

$3w + w = 24 + 8$

$4w = 32$

$\dfrac{4w}{4} = \dfrac{32}{4}$

$w = 8$

11. $6y + 3(y + 1) = y - 2(y - 3) + 7$

$6y + 3y + 3 = y - 2y + 6 + 7$

$9y + 3 = -y + 13$

$9y + y = 13 - 3$

$10y = 10$

$y = 1$

12. $-3y - 3 = 2$

$-3y = 2 + 3$

$-3y = 5$

$\dfrac{-3y}{-3} = \dfrac{5}{-3}$

$y = -1\dfrac{2}{3}$

13. $-3(m - 1) + m = 2(m - 1) - 3$

$-3m + 3 + m = 2m - 2 - 3$

$2m + 3 = 2m - 5$

$-2m - 2m = -5 - 3$

$-4m = -8$

$\dfrac{-4m}{-4} = \dfrac{-8}{-4}$

$m = 2$

14. $9x - 5(3x + 4) = 14 + 2(x - 1)$

$9x - 15x - 20 = 14 + 2x - 2$

$-6x - 20 = 12 + 2x$

$-6x - 2x = 12 + 20$

$-8x = 32$

$\dfrac{-8x}{-8} = \dfrac{32}{-8}$

$x = -4$

15. $2(t - 3) - 2(3t - 4) = 2 + 4(t - 6)$

$2t - 6 - 6t + 8 = 2 + 4t - 24$

$-4t + 2 = 4t - 22$

$-4t - 4t = -2 - 22$

$-8t = -24$

$\dfrac{-8t}{-8} = \dfrac{-24}{-8}$

$t = 3$

Answers to Assignments

I   1. Conditional equation   3. Identity

II   5. $x = 9$ 7. $y = -7$ 9. $y = 5$

11. $w = -\dfrac{5}{2}$ 13. $x = -2$ 15. $z = -\dfrac{1}{3}$

Unit 18

Solutions to Drill Exercises

I   1. $x + 17$ 2. $6n$ 3. $4m - 9$ 4. $x + 9x^2$ 5. $11z + 3$

II   6.   *Step 1*   Let n = the number.

   Step 2   6n =   3n + 9

            6n – 3n =   9

                3n =   9

   Step 3   n =   3

   Step 4   6(3) = 18 = 3(3) + 9

   Step 5   Hence, 3 is correct.

            ∴ the number is 3.

7.   *Step 1*   Let x = the number.

   Step 2   4x – 11 =   9

                4x =   20

   Step 3   x =   5

   Step 4   Check   4(5) – 11 = 20 – 11 = 9

   Step 5   Hence, 5 is correct.

            ∴ the number is 5.

8.   *Step 1*   Let x = the length of one piece.

            Then x – 68 = the length of the other piece.

   Step 2   x + (x – 68) = 200

   Step 3   2x – 68 = 200

                2x = 268

   Step 4   x = 134

            x – 68 = 66

   Step 5   Check: 134 + 66 = 200 cm

   Step 6   Hence, the pieces are 134 cm and 66 cm in length.

9.   *Step 1*   Let m = the cost of the lot.

            Then 5m – 10 000 = the cost of the house.

   Step 2   m + (5m – 10 000) = 98 000

   Step 3   6m – 10 000   =   98 000

                6m =   108 000

                m =   18 000

   Step 4   5m – 10 000   =   5(18 000) – 10 000

                    =   90 000 – 10 000

                    =   80 000

   Step 5   Check: 18 000 + 80 000 = $98 000

   Step 6   Hence, the house is worth $80 000.

10.   *Step 1*   Let m = the number of Type A batteries.

            Then 32 – m = the number of Type B batteries.

   Step 2   50(m) + 65(32 – m)   =   1780

                    –15m =   –300

                    m =   20

   Step 3   32 – m = 32 – 20 = 12

II   *Step 4*   Check: 50(20) + 65(12)

                                   =   10 000 + 780

                                   =   1780

          Step 5   Hence, he purchased 20 Type A batteries and 12 Type B batteries.

11.   *Step 1*   Let n = the number of nickles.

                           2n = the number of quarters.

                   17 − (2n + n) = the number of dimes

          Step 2   5n + 10(17 − 3n) + 25(2n) = 270

                              5n + 170 − 30n + 50n = 270

                                              25n = 100

          Step 3   n = 4

          Step 4   2(4) = 8   17 − 3(4) = 17 − 12 = 5

          Step 5   Check: 4 + 8 + 5 = 17

          Step 6   Hence, there are 4 nickels, 5 dimes and 8 quarters.

Answers to Assignments

1.   14   3.   18, 4, 5   5.   6, 33   7.   10 dimes, 1 nickel, 3 pennies

9.   X costs $244, Y costs $532, Z costs $200

11.   first circuit carries 22 amperes, second 38, third 10, fourth 19   13.   50 mm

---

Unit 19

Solutions to Drill Exercises

I   1. monomial   2. trinomial   3. monomial   4. binomial   5. monomial

II   6. $(3m^2n^3)(2mn) = 6m^3n^4$ 7. $(2x^{-2}y^4)(-x^2y)$ = $-2x^0y^5$

                                                                                      = $-2y^5$

     8. $(-2x^2yz)(-x^2yz^2)(3x^{-4}yz^{-1})$ = $6x^0y^3\,z^2$ 9. $3(p^3q + r^3) = 3p^3q + 3r^3$

                                              = $6y^3z^2$

     10. $-ab(5a^3c^2 + 4b^3c^4) = -5a^4bc^2 - 4ab^4c^4$

     11. $4mn(5m + 2n - 6)$ = $(4mn)(5m) + (4mn)(2n) - (4mn)(6)$

                              = $20m^2n + 8mn^2 - 24mn$

     12. $-5x^2y(2x^3 + y - z - 4)$ = $(-5x^2y)(2x^3) + (-5x^2y)(y) - (-5x^2y)(z) - (-5x^2y)(4)$

                                      = $-10x^5y - 5x^2y^2 + 5x^2yz + 20x^2y$

II   13. $(x + 4)(2x^2 + 3x + 2)$ = $x(2x^2 + 3x + 2) + 4(2x^2 + 3x + 2)$

$$= 2x^3 + 3x^2 + 2x + 8x^2 + 12x + 8$$

$$= 2x^3 + 11x^2 + 14x + 8$$

14. $(3m - 2n)(4m - n)$ = $12m^2 - 3mn - 8mn + 2n^2$

$$= 12m^2 - 11mn + 2n^2$$

15. $2x(x - 3)(x + 2)$ = $2x(x^2 + 2x - 3x - 6)$

$$= 2x(x^2 - x - 6)$$

$$= 2x^3 - 2x^2 - 12x$$

16. $(x - 2)(x + 3)(x - 1)$ = $(x^2 + 3x - 2x - 6)(x - 1)$

$$= (x^2 + x - 6)(x - 1)$$

$$= (x^2 + x - 6)(x) - (x^2 + x - 6)(1)$$

$$= (x^3 + x^2 - 6x) - (x^2 + x - 6)$$

$$= x^3 + x^2 - 6x - x^2 - x + 6$$

$$= x^3 - 7x + 6$$

17. $(a - b)(a^2 - 2ab + b^2)$ = $a(a^2 - 2ab + b^2) - b(a^2 - 2ab + b^2)$

$$= a^3 - 2a^2b + ab^2 - ba^2 + 2ab^2 - b^3$$

$$= a^3 - 3a^2b + 3ab^2 - b^3$$

18. $(3m + 4)(m^2 - m - 1)$ = $3m(m^2 - m - 1) + 4(m^2 - m - 1)$

$$= 3m^3 - 3m^2 - 3m + 4m^2 - 4m - 4$$

$$= 3m^3 + m^2 - 7m - 4$$

19. $(x^2 + 2x - 3)(2x^2 - x - 5)$ = $x^2(2x^2 - x - 5) + 2x(2x^2 - x - 5) - 3(2x^2 - x - 5)$

$$= 2x^4 - x^3 - 5x^2 + 4x^3 - 2x^2 - 10x - 6x^2 + 3x + 15$$

$$= 2x^4 + 3x^3 - 13x^2 - 7x + 15$$

III   20.
$$
\begin{array}{r}
3x^2 + 2xy - y^2 \\
x + 4y \\
\hline
3x^3 + 2x^2y - xy^2 \\
12x^2y + 8xy^2 + 4y^3 \\
\hline
3x^3 + 14x^2y + 7xy^2 - 4y^3
\end{array}
$$

Hence, $(x + 4y)(3x^2 + 2xy - y^2) = 3x^3 + 14x^2y + 7xy^2 - 4y^3$.

21.
$$
\begin{array}{r}
m - 3n - 7 \\
2m - n \\
\hline
2m^2 - 6mn - 14m \\
- mn + 3n^2 + 7n \\
\hline
2m^2 - 7mn - 14m + 3n^2 + 7n
\end{array}
$$

Hence, $(2m - n)(m - 3n - 7) = 2m^2 - 7mn - 14m + 3n^2 + 7n$

III   22. $\begin{aligned} 5a^2 + 2a - 3 \\ a^2 - 3a + 2 \end{aligned}$

$$5a^4 + 2a^3 - 3a^2$$
$$- 15a^3 - 6a^2 + 9a$$
$$10a^2 + 4a - 6$$
$$\overline{5a^4 - 13a^3 + a^2 + 13a - 6}$$

Hence, $(a^2 - 3a + 2)(5a^2 + 2a - 3) = 5a^4 - 13a^3 + a^2 + 13a - 6$

IV   23. $(y + 8)^2 = y^2 + 2(8y) + 8^2$

$= y^2 + 16y + 64$

24. $(x + 4)(x + 4) = (x + 4)^2$

$= x^2 + 2(4x) + 4^2$

$= x^2 + 8x + 16$

25. $(a + 4)(a - 4) = a^2 - 4^2$

$= a^2 - 16$

26. $(x - 4)^2 = x^2 + 2(-4x) + (-4)^2$

$= x^2 - 8x + 16$

27. $(3a + 3b)^2 = (3a)^2 + 2(9ab) + (3b)^2$

$= 9a^2 + 18ab + 9b^2$

28. $(2m + 3n)(2m - 3n) = (2m)^2 - (3n)^2$

$= 4m^2 - 9n^2$

29. $(6x^2 + 2y)(6x^2 - 2y) = (6x^2)^2 - (2y)^2$

$= 36x^4 - 4y^2$

30. $(5y - 2z)^2 = (5y)^2 + 2(-10yz) + (-2z)^2$

$= 25y^2 - 20yz + 4z^2$

Answers to Assignments

I   1. $-7x^3$

3. $p^2 + 6p + 9$

5. $6mn^2 - 6m$

7. $-4b^2c^2$

9. $-3x^5y^2z + 8x^2yz^4$

11. $15x^2 - 18xy + 3y^2$

13. $4y^3 + 11y^2 + 9y + 6$

15. $16a^2 - 4b^2$

17. $2x^3 - 3x^2y + 3xy^2 - 2y^3$

19. $m^4 - 9n^4$

21. $4x^2 + 12xy + 9y^2$

II   23. $10x^7 + 9x^5 + 4x^4 - 29x^3 + 6x^2 + 12x - 8$

25. $x^3 + 1$

Unit 20

Solutions to Drill Exercises

I   1. $\dfrac{24x^3y^2z^2}{4xyz^2} = 6x^{3-1}y^{2-1}z^{2-2}$

$= 6x^2y^1z^0$

$= 6x^2y$

2. $(-16a^4b^6c) \div (-8a^2b^2c) = 2a^{4-2}b^{6-2}c^{1-1}$

$= 2a^2b^4c^0$

$= 2a^2b^4$

I   3. $\dfrac{-12x^4yz^3}{3x^2y^4z}$ $= -4x^{4-2}y^{1-4}z^{3-1}$

                  $= -4x^2y^{-3}z^2$

                  $= \dfrac{-4x^2z^2}{y^3}$

4. $18r^3s^2t \div (-4r^5st^2)$ $= -\dfrac{9}{2}r^{3-5}s^{2-1}t^{1-2}$

                             $= \dfrac{-9r^{-2}s^1t^{-1}}{2}$

                             $= \dfrac{-9s}{2r^2t}$

5. $\dfrac{4a^3b^2 + 16ab - a^2}{2a^2b}$ $= \dfrac{4a^3b^2}{2a^2b} + \dfrac{16ab}{2a^2b} - \dfrac{a^2}{2a^2b}$

                           $= 2ab + \dfrac{8}{a} - \dfrac{1}{2b}$

6. $(3x^3 + 16xy^2 - 12x^4yz^4) \div (-2x^2yz)$

   $= \dfrac{3x^3 + 16xy^2 - 12x^4yz^4}{-2x^2yz}$

   $= \dfrac{3x^3}{-2x^2yz} + \dfrac{16xy^2}{-2x^2yz} - \dfrac{12x^4yz^4}{-2x^2yz}$

   $= -\dfrac{3x}{2yz} - \dfrac{8y}{xz} + 6x^2z^3$

7. $\dfrac{-4ab^3 - 3a^2bc - 12a^3b^2c^4}{-2ab^2c^3}$ $= \dfrac{-4ab^3}{-2ab^2c^3} - \dfrac{3a^2bc}{-2ab^2c^3} - \dfrac{12a^3b^2c^4}{-2ab^2c^3}$

                                 $= \dfrac{2b}{c^3} + \dfrac{3a}{2bc^2} + 6a^2c$

8. $(-14m^3n^5 - 42mn^2 + 7m^4n) \div (7mn)$

   $= \dfrac{-14m^3n^5 - 42mn^2 + 7m^4n}{7mn}$

   $= \dfrac{-14m^3n^5}{7mn} - \dfrac{42mn^2}{7mn} + \dfrac{7m^4n}{7mn}$

   $= -2m^2n^4 - 6n + m^3$

II   9. $9x + 3y = 3(3x + y)$ 10. $8ab^2 - 4b = 4b(2ab - 1)$

11. $10n^6 + 5n^5 - 15n^2 = 5n^2(2n^4 + n^3 - 3)$ 12. $3x^3y^2z - 6x^2z + 15y = 3(x^3y^2z - 2x^2z + 5y)$

13. $21p^4q^5 - 7p^3q^3 - 14pq^4 = 7pq^3(3p^3q^2 - p^2 - 2q)$

14. $3x^5yz^2 + 9x^4y^3z - 12x^3yz^2 = 3x^3yz(x^2z + 3xy^2 - 4z)$

15. $36s^4r^2 + 42s^3r = 6s^3r(6sr + 7)$ 16. $7m^3n^4p + 21m^2n^3p^2 = 7m^2n^3p(mn + 3p)$

Answers to Assignments

I   1. $\dfrac{-2r^2t}{3s}$ 3. $\dfrac{-1}{3xy}$ 5. $-\dfrac{3}{2yz} + \dfrac{8y^2}{xz} + \dfrac{9x}{2}$ 7. $-6u + 2u^6v^2 + 7v$

II   9. $5(m - 5n)$ 11. $3a^3b(2a^2b - a + 3b^2)$ 13. $8a(-4 + a^2)$ 15. $4ab(2a^2 - 3b)$

Unit 21

Solutions to Drill Exercises

I   (a)   1. $\dfrac{4b}{3a + 2} = (4b) \div (3a + 2)$ 2. $\dfrac{2x^2 + x + 1}{x^2 - 1} = (2x^2 + x + 1) \div (x^2 - 1)$

3. $\dfrac{7}{n - 1} = 7 \div (n - 1)$

(b)   4. $(m^2 - 3) \div (2m + 7) = \dfrac{m^2 - 3}{2m + 7}$

5. $y^2 - 3x^3y \div x + 3y = y^2 - \dfrac{3x^3y}{x} + 3y$

6. $p^2 + 6p \div (2p - 1) = p^2 + \dfrac{6p}{2p - 1}$

II   7. If $y = 2$, $\dfrac{4y - 3}{7y} = \dfrac{4(2) - 3}{7(2)} = \dfrac{8 - 3}{14} = \dfrac{5}{14}$

8. If $x = 5, y = 3$, $\dfrac{2x + y^2}{3(x - y)} = \dfrac{2(5) + 3^2}{3(5 - 3)} = \dfrac{10 + 9}{3(2)} = \dfrac{19}{6}$ or $3\dfrac{1}{6}$

9. If $m = 2, n = 3$, $\dfrac{5m^2 + 4n}{(m - 2)(n - 3)} = \dfrac{5(4) + 4(3)}{(0)(0)} = \dfrac{32}{0}$ undefined.

III   10. $\dfrac{4}{3x}$, $x \neq 0$ 11. $\dfrac{5y + 2}{y + 3}$, $y \neq -3$ 12. $\dfrac{3a - 1}{b(a - 2)}$, $b \neq 0$, $a \neq 2$

13. $\dfrac{2a^2 + 3b - 4}{3(a + 2)(b - 3)}$, $a \neq -2$, $b \neq 3$ 14. $\dfrac{2m + 3}{m^2 - 4}$, $m \neq 2$, $m \neq -2$

IV   15. $\dfrac{7}{8} = \dfrac{7 \cdot 4}{8 \cdot 4} = \dfrac{28}{32}$ 16. $\dfrac{3m}{5n} = \dfrac{(3m)(4n^2)}{(5n)(4n^2)} = \dfrac{12mn^2}{20n^3}$

17. $\dfrac{a + 2}{b} = \dfrac{(a + 2)(ab)}{b(ab)} = \dfrac{(a + 2)(ab)}{ab^2}$ or $\dfrac{a^2b + 2ab}{ab^2}$

18. $\dfrac{5}{x - 2} = \dfrac{5(x + 1)}{(x - 2)(x + 1)}$ or $\dfrac{5x + 5}{(x - 2)(x + 1)}$

V   19. Step 1. $21 = 3 \cdot 7$ and $49 = 7 \cdot 7$

Step 2. $\dfrac{21}{49} = \dfrac{3 \cdot \cancel{7}^{\,1}}{7 \cdot \cancel{7}_{\,1}} = \dfrac{3}{7}$

20. Step 1. $27x^2y^4 = 3(9x^2y^4)$ and $36x^4y^7 = (4x^2y^3)(9x^2y^4)$

Step 2. $\dfrac{27x^2y^4}{36x^4y^7} = \dfrac{3\cancel{(9x^2y^4)}^{\,1}}{(4x^2y^3)\cancel{(9x^2y^4)}_{\,1}} = \dfrac{3}{4x^2y^3}$

21. Step 1. $6a^2 - 8b = 2(3a^2 - 4b)$ and $2a^2 + 14b = 2(a^2 + 7b)$

Step 2. $\dfrac{6a^2 - 8b}{2a^2 + 14b} = \dfrac{\cancel{2}^{\,1}(3a^2 - 4b)}{\cancel{2}_{\,1}(a^2 + 7b)} = \dfrac{3a^2 - 4b}{a^2 + 7b}$

V   22.   Step 1. $3m + 6n = 3(m + 2n)$ and $4m + 8n = 4(m + 2n)$

Step 2. $\dfrac{3m + 6n}{4m + 8n} = \dfrac{3\cancel{(m + 2n)}^{1}}{4\cancel{(m + 2n)}_{1}} = \dfrac{3}{4}$

23.   Step 1. $7x^3 - 14x^2 = 7x^2(x - 2)$

Step 2. $\dfrac{7x}{7x^3 - 14x^2} = \dfrac{7x}{7x^2(x - 2)} = \dfrac{\cancel{7x}^{1}}{\cancel{7x}_{1} \cdot x(x - 2)} = \dfrac{1}{x(x - 2)}$

24.   Step 1. $3b^4 - 9b^3 = 3b^3(b - 3)$ and $6b^5 + 15b^6 = 3b^5(2 + 5b)$

Step 2. $\dfrac{3b^4 - 9b^3}{6b^5 + 15b^6} = \dfrac{3b^3(b - 3)}{3b^5(2 + 5b)} = \dfrac{\cancel{3b^3}^{1}(b - 3)}{(\cancel{3b^3})_{1}(b^2)(2 + 5b)} = \dfrac{b - 3}{b^2(2 + 5b)}$

VI   25. $\dfrac{-5}{13} = \dfrac{5}{-13}$ is true because if a fraction has one negative sign, it can be placed in front of either the numerator,

the fraction bar, or the denominator.

26.   False. $\dfrac{-a^2b}{-b^2} = \dfrac{a^2b}{b^2}$

27.   False. $\dfrac{x^2 - 1}{-3x^2} = \dfrac{-(x^2 - 1)}{3x^2} = \dfrac{-x^2 + 1}{3x^2}$

28. $-\dfrac{-(2m + n)}{-(m + 2n)} = -\dfrac{2m + n}{m + 2n}$ is true because any two of the three signs of a fraction can be changed

without changing the value of the fraction.

VII   29.   Step 1. $2 - m = -(m - 2)$

Step 2. $\dfrac{m - 2}{2 - m} = \dfrac{\cancel{m - 2}^{1}}{-(\cancel{m - 2})_{1}} = -1$

30.   Step 1. $4 - y = -(y - 4)$

Step 2. $\dfrac{3(y + 3)(y - 4)}{9(4 - y)(y - 3)} = \dfrac{\cancel{3}(y + 3)\cancel{(y - 4)}^{1}}{-\cancel{(y - 4)}_{1}(\cancel{3})(3)(y - 3)} = -\dfrac{y + 3}{3(y - 3)}$

Answers to Assignments

I   (a)   1. $(x^2 + 1) \div (x - 1)$ (b)   3. $\dfrac{m^2 + 3m - 1}{m} + 2$

II   5.   undefined   7.   undefined

III   9. $x \neq 5$ or $x \neq -5$

IV   11. $\dfrac{12}{21}$ 13. $\dfrac{5x}{x(x^2 + 1)}$

V   15. $\dfrac{2}{3}$ 17. $\dfrac{1}{m + n}$ 19. $\dfrac{5y - 1}{7}$

VI   21.   true   23.   false $-\dfrac{2m - n}{mn} = \dfrac{-2m + n}{mn}$

VII   25. -1 27. $\dfrac{-(a + 2)}{2(a + 3)}$

Unit 22

Solutions to Drill Exercises

I   1. $\dfrac{2}{7} \cdot \dfrac{m}{n} = \dfrac{2m}{7n}$

2. $\left(\dfrac{3x}{5y}\right)\left(\dfrac{x}{2y}\right) = \dfrac{3x^2}{10y^2}$

3. $\dfrac{1}{b} \cdot \dfrac{1}{4} = \dfrac{1}{4b}$

4. $\dfrac{5b^2}{6a} \cdot \dfrac{1}{a^4} = \dfrac{5b^2}{6a^5}$

5. $\left(\dfrac{7m^2}{2n^4}\right)\left(\dfrac{3m}{4n^2}\right) = \dfrac{21m^3}{8n^6}$

6. $\left(\dfrac{3}{y^5}\right)\left(\dfrac{-2x}{y}\right) = \dfrac{-6x}{y^6}$ or $-\dfrac{6x}{y^6}$

7. $9 \cdot \dfrac{2V}{p} = \dfrac{9}{1} \cdot \dfrac{2V}{p} = \dfrac{18V}{p}$

8. $(2y)\left(\dfrac{5xy}{3z}\right) = \left(\dfrac{2y}{1}\right)\left(\dfrac{5xy}{3z}\right) = \dfrac{10xy^2}{3z}$

9. $\dfrac{n-2}{2n+1} \cdot 3n^3 = \dfrac{n-2}{2n+1} \cdot \dfrac{3n^3}{1} = \dfrac{3n^3(n-2)}{2n+1}$ or $\dfrac{3n^4 - 6n^3}{2n+1}$

II   10. $\dfrac{6}{11} = 6 \cdot \dfrac{1}{11}$

11. $\dfrac{2x}{-3y^2} = (2x) \cdot \dfrac{1}{-3y^2}$ or $(2x)\left(-\dfrac{1}{3y^2}\right)$

12. $\dfrac{a^3 - 5a}{3b^2} = (a^3 - 5a) \cdot \dfrac{1}{3b^2}$

III   13. $\dfrac{13}{15} \cdot \dfrac{30}{39} = \dfrac{13}{15} \cdot \dfrac{2 \cdot 15}{3 \cdot 13} = \dfrac{2}{3}$

14. $\dfrac{4a^3}{3b^2} \cdot \dfrac{15b}{16a} = \dfrac{4 \cdot a \cdot a^2}{3 \cdot b \cdot b} \cdot \dfrac{3 \cdot 5 \cdot b}{4 \cdot 4 \cdot a} = \dfrac{5a^2}{4b}$

15. $\dfrac{3x}{5x-5} \cdot \dfrac{2x-2}{9x^3} = \dfrac{3x}{5(x-1)} \cdot \dfrac{2(x-1)}{(3x)(3x^2)} = \dfrac{2}{15x^2}$

16. $\dfrac{-x^2y^2}{14z^2} \cdot \dfrac{-7x^3z^3}{4y^4} = \dfrac{-x^2y^2}{2(7z^2)} \cdot \dfrac{-x^3z(7z^2)}{(4y^2)(y^2)}$

$= \dfrac{x^5z}{8y^2}$

IV   17. The reciprocal of $\dfrac{19}{23}$ is $\dfrac{23}{19}$

18. The reciprocal of $\dfrac{-2x^3}{7}$ is $\dfrac{7}{-2x^3}$ or $-\dfrac{7}{2x^3}$

19. The reciprocal of $\dfrac{a^2b^3}{3a^2 + 2b^2}$ is $\dfrac{3a^2 + 2b^2}{a^2b^3}$

20. The reciprocal of 15 is $\dfrac{1}{15}$

21. The reciprocal of $m^2n + mn + 2$ is $\dfrac{1}{m^2n + mn + 2}$

V   22. $\dfrac{a}{7} \div \dfrac{2}{b} = \dfrac{a}{7} \cdot \dfrac{b}{2} = \dfrac{ab}{14}$

23. $\dfrac{x-3}{5y} \div \dfrac{y+7}{6x} = \dfrac{x-3}{5y} \cdot \dfrac{6x}{y+7} = \dfrac{6x(x-3)}{5y(y+7)}$ or $\dfrac{6x^2 - 18x}{5y^2 + 35y}$

V 24. $\dfrac{\dfrac{m^3}{3n^2}}{\dfrac{n+1}{m-1}}$ $=$ $\dfrac{m^3}{3n^2} \div \dfrac{n+1}{m-1}$

$\qquad\qquad\qquad$ 25. $\dfrac{4a}{3b} \div c^3 = \dfrac{4a}{3b} \cdot \dfrac{1}{c^3} = \dfrac{4a}{3bc^3}$

$\qquad\qquad = \dfrac{m^3}{3n^2} \cdot \dfrac{m-1}{n+1}$

$\qquad\qquad = \dfrac{m^3(m-1)}{3n^2(n+1)} \quad \text{or} \quad \dfrac{m^4-m^3}{3n^3+3n^2}$

26. $\dfrac{x^2-2x}{y^3} \div (y+5) = \dfrac{x^2-2x}{y^3} \cdot \dfrac{1}{y+5}$

$\qquad\qquad\qquad\qquad = \dfrac{x^2-2x}{y^3(y+5)} \quad \text{or} \quad \dfrac{x^2-2x}{y^4+5y^3}$

27. $\dfrac{3m+6}{m^3} \div \dfrac{m+2}{m^2} = \dfrac{3m+6}{m^3} \cdot \dfrac{m^2}{m+2}$

$\qquad\qquad\qquad\qquad = \dfrac{3(m+2)}{m \cdot m^2} \cdot \dfrac{m^2}{m+2}$

$\qquad\qquad\qquad\qquad = \dfrac{3}{m}$

28. $\dfrac{2(x+3)}{x^2y} \div \dfrac{y}{6(x+3)} = \dfrac{2(x+3)}{x^2y} \cdot \dfrac{6(x+3)}{y}$

$\qquad\qquad\qquad\qquad\qquad = \dfrac{12(x+3)^2}{x^2y^2}$

VI 29. $\dfrac{6x^2+3x}{6x} \cdot \dfrac{10x^2}{12x+6} = \dfrac{3x(2x+1)}{2(3x)} \cdot \dfrac{2 \cdot 5x^2}{6(2x+1)}$

$\qquad\qquad\qquad\qquad\qquad = \dfrac{5x^2}{6}$

Answers to Assignments

I 1. $\dfrac{5x}{9y}$ $\qquad\qquad$ 3. $\dfrac{9m}{4n}$ $\qquad\qquad$ 5. $\dfrac{2a^4}{15b^7}$ $\qquad\qquad$ 7. $\dfrac{5z}{4x^2y}$

$\quad$ 9. $-\dfrac{4}{x}$ $\qquad\qquad$ 11. $\dfrac{x}{4}$ $\qquad\qquad$ 13. $\dfrac{2}{15ab}$

II 15. $\dfrac{5x^3}{2y^4}$ $\qquad\qquad$ 17. $\dfrac{x^2}{2zy^3}$ $\qquad\qquad$ 19. $\dfrac{3m}{(n-1)^2}$

Unit 23

Solutions to Drill Exercises

I 1. $\dfrac{3V}{4} + \dfrac{1}{4} = \dfrac{3V+1}{4}$ $\qquad\qquad$ 2. $\dfrac{y^4}{x} - \dfrac{2}{x} = \dfrac{y^4-2}{x}$

$\quad$ 3. $\dfrac{7}{m^3} - \dfrac{5}{m^3} = \dfrac{7-5}{m^3} = \dfrac{2}{m^3}$ $\qquad\qquad$ 4. $\dfrac{2b}{a+2} + \dfrac{3}{a+2} = \dfrac{2b+3}{a+2}$

$\quad$ 5. $\dfrac{3z^2}{x^2y^2} + \dfrac{6z^2}{x^2y^2} = \dfrac{3z^2+6z^2}{x^2y^2} = \dfrac{9z^2}{x^2y^2}$

I 6. $\dfrac{T^2S}{2T - 1} - \dfrac{-TS^2}{2T - 1} = \dfrac{T^2S - (-TS^2)}{2T - 1} = \dfrac{T^2S + TS^2}{2T - 1}$

7. $\dfrac{5y + 4}{7y^2} + \dfrac{2y - 3}{7y^2} = \dfrac{5y + 4 + 2y - 3}{7y^2} = \dfrac{7y + 1}{7y^2}$

8. $\dfrac{9x - 2}{3b} - \dfrac{x + 3}{3b} = \dfrac{9x - 2}{3b} - \dfrac{(x + 3)}{3b}$

$= \dfrac{9x - 2 - (x + 3)}{3b}$

$= \dfrac{9x - 2 - x - 3}{3b}$

$= \dfrac{8x - 5}{3b}$

9. $\dfrac{6n^3 - n}{5m} - \dfrac{2n^3 - 3n}{5m} = \dfrac{6n^3 - n}{5m} - \dfrac{(2n^3 - 3n)}{5m}$

$= \dfrac{6n^3 - n - (2n^3 - 3n)}{5m}$

$= \dfrac{6n^3 - n - 2n^3 + 3n}{5m}$

$= \dfrac{4n^3 + 2n}{5m}$

II 10.

```
2 | 600
2 | 300
2 | 150
3 | 75
5 | 25
5 | 5
    1
```

11.

```
2 | 162
3 | 81
3 | 27
3 | 9
3 | 3
    1
```

$600 = 2^3 \cdot 3 \cdot 5^2$

$162x^3y = 2 \cdot 3^4x^3y$

12. $9y^3 - 36y^2 = 9y^2(y - 4) = 3^2y^2(y - 4)$

13. $4x^3 + 4x^2 + 8x = 4x(x^2 + x + 2) = 2^2x(x^2 + x + 2)$

14. $24m^4 - 56m^3 = 8m^3(3m - 7) = 2^3m^3(3m - 7)$

15. $42z^2 + 21z^3 = 21z^2(2 + z) = 3 \cdot 7 \cdot z^2(2 + z)$

III 16. The LCD of $\dfrac{3}{5}$, $\dfrac{2}{7}$ and $\dfrac{3}{35}$ is 35, since 5 and 7 are factors of 35.

17. The LCD of $\dfrac{3m^2}{n}$ and $\dfrac{m}{v}$ is nv by inspection.

18. The LCD of $\dfrac{4y}{x + 3}$ and $\dfrac{3}{x - 2}$ is $(x + 3)(x - 2)$.

19. The LCD of $\dfrac{z^2}{2x}$ and $\dfrac{4}{3y}$ is 6xy.

20. The LCD of $\dfrac{x^2}{(x + 3)^2}$ and $\dfrac{2x}{x + 3}$ is $(x + 3)^2$.

21. The LCD of $\dfrac{4}{3a}$ and $\dfrac{2}{9a^2}$ is $9a^2$.

III 22.
$$6 = 2 \cdot 3$$
$$18 = 2 \cdot 3^2$$
$$63 = 3^2 \cdot 7$$
$$\text{LCD} = 2 \cdot 3^2 \cdot 7$$
$$= 126$$

23.
$$16ab^3c^2 = 2^4 \cdot a \cdot b^3 \cdot c^2$$
$$28a^2c^4 = 2^2 \cdot 7 \cdot a^2 \cdot c^4$$
$$\text{LCD} = 2^4 \cdot 7 \cdot a^2 \cdot b^3 \cdot c^4$$
$$= 112a^2b^3c^4$$

24.
$$36x^2y^2 = 2^2 \cdot 3^2 \cdot x^2 \cdot y^2$$
$$8x^4 - 24x^3 = 8x^3(x - 3)$$
$$= 2^3x^3(x - 3)$$
$$\text{LCD} = 2^3 \cdot 3^2 \cdot x^3 \cdot y^2(x - 3)$$
$$= 72x^3y^2(x - 3)$$

IV 25. Since $15 = 3 \cdot 5$, $18 = 2 \cdot 3^2$ and $20 = 2^2 \cdot 5$, the LCD $= 2^2 \cdot 3^2 \cdot 5 = 180$.

$$\frac{8}{15} + \frac{5}{18} - \frac{7}{20} = \frac{8 \cdot 12}{15 \cdot 12} + \frac{5 \cdot 10}{18 \cdot 10} - \frac{7 \cdot 9}{20 \cdot 9}$$

$$= \frac{96}{180} + \frac{50}{180} - \frac{63}{180}$$

$$= \frac{96 + 50 - 63}{180}$$

$$= \frac{83}{180}$$

26. LCD $x^3 \cdot$ (by inspection).

$$\frac{7}{x} - \frac{2}{x^3} = \frac{7\,x^2}{x\,x^2} - \frac{2}{x^3}$$

$$= \frac{7x^2}{x^3} - \frac{2}{x^3}$$

$$= \frac{7x^2 - 2}{x^3}$$

27. LCD $= wxy$

$$\frac{2y}{wx} + \frac{3w}{y} = \frac{(2y)y}{(wx)y} + \frac{(3w)(wx)}{y(wx)}$$

$$= \frac{2y^2}{wxy} + \frac{3w^2x}{wxy}$$

$$= \frac{2y^2 + 3w^2x}{wxy}$$

28. LCD $= xyz$

$$\frac{2}{xy} + \frac{3}{xz} + \frac{4}{yz} = \frac{2(z)}{(xy)(z)} + \frac{3(y)}{(xz)y} + \frac{4(x)}{(yz)x}$$

$$= \frac{2z}{xyz} + \frac{3y}{xyz} + \frac{4x}{xyz}$$

$$= \frac{2z + 3y + 4x}{xyz}$$

IV 29. Since $10mn^2 = 2 \cdot 5 \cdot m \cdot n^2$ and $35m^4n = 5 \cdot 7 \cdot m^4n$, the LCD $= 2 \cdot 5 \cdot 7 \cdot m^4 \cdot n^2$

$$= 70m^4n^2$$

$$\frac{3p^2}{10mn^2} - \frac{-2p}{35m^4n} = \frac{(3p^2)(7m^3)}{(10mn^2)(7m^3)} - \frac{(-2p)(2n)}{(35m^4n)(2n)}$$

$$= \frac{21m^3p^2}{70m^4n^2} - \frac{(-4np)}{70m^4n^2}$$

$$= \frac{21m^3p^2 - (-4np)}{70m^4n^2}$$

$$= \frac{21m^3p^2 + 4np}{70m^4n^2}$$

30.

$$\text{LCD} = (m+1)(n-2)$$

$$\frac{2}{m+1} + \frac{3}{n-2} = \frac{2(n-2)}{(m+1)(n-2)} + \frac{3(m+1)}{(n-2)(m+1)}$$

$$= \frac{2(n-2) + 3(m+1)}{(m+1)(n-2)}$$

$$= \frac{2n - 4 + 3m + 3}{(m+1)(n-2)}$$

$$= \frac{2n + 3m - 1}{(m+1)(n-2)}$$

31.

$$\text{Since } 4x^2 - 8x = 4x(x-2) = 2^2x(x-2)$$

$$\text{and } 16x^2 = 2^4x^2,$$

$$\text{the LCD} = 2^4x^2(x-2)$$

$$= 16x^2(x-2)$$

$$\frac{7y^2}{4x^2 - 8x} - \frac{3}{16x^2} = \frac{7y^2}{4x(x-2)} - \frac{3}{16x^2}$$

$$= \frac{7y^2(4x)}{4x(x-2)(4x)} - \frac{3(x-2)}{16x^2(x-2)}$$

$$= \frac{28xy^2}{16x^2(x-2)} - \frac{(3x-6)}{16x^2(x-2)}$$

$$= \frac{28xy^2 - (3x-6)}{16x^2(x-2)}$$

$$= \frac{28xy^2 - 3x + 6}{16x^2(x-2)}$$

Answers to Assignments

1. $\dfrac{3 - 2x}{8}$

3. $\dfrac{3bc + 14ad}{21ab}$

5. $\dfrac{p - R}{(p+1)(R+1)}$

7. $\dfrac{5 + xy - 2y}{(x-2)^2}$

9. $\dfrac{4 + n}{m - 3}$

11. $\dfrac{V^2w}{xy^3}$

13. $\dfrac{16x^4 - x + 4}{72x^3(x-4)}$

15. 1

17. $\dfrac{21z^2 + 5x^7y}{3x^7yz}$

19. $\dfrac{-28bd^2 + 15ac^2}{120a^2c^2d^3}$

Unit 24

Solutions to Drill Exercises

I 1. LCD = 4

$$4 \cdot \frac{1}{4} m = 4 \cdot 7$$

$$m = 28$$

Check:

L.H.S. $= \frac{1}{4} \cdot 28 = 7$

R.H.S. $= 7$

Hence, the solution is m = 28.

2. LCD = 6

$$6 \cdot \frac{2}{3} p = 6 \cdot \frac{1}{2}$$

$$4p = 3$$

$$p = \frac{3}{4}$$

Check:

L.H.S. $= \frac{2}{3} \cdot \frac{3}{4} = \frac{1}{2}$

R.H.S. $= \frac{1}{2}$

Hence, the solution is p $= \frac{3}{4}$.

3. LCD = 14

$$14 \left(\frac{5}{7} V - \frac{1}{14} \right) = 14 \cdot 0$$

$$14 \cdot \frac{5}{7} V - 14 \cdot \frac{1}{14} = 14 \cdot 0$$

$$10V - 1 = 0$$

$$10V = 1$$

$$V = \frac{1}{10}$$

Check:

L.H.S. $= \frac{5}{7} \cdot \frac{1}{10} - \frac{1}{14} = \frac{1}{14} - \frac{1}{14} = 0$

R.H.S. $= 0$

Hence, V $= \frac{1}{10}$ is the solution.

4. LCD = 24

$$24 \cdot \frac{5}{8} = 24 \left(\frac{2}{3} R - \frac{1}{4} \right)$$

$$24 \cdot \frac{5}{8} = 24 \cdot \frac{2}{3} R - 24 \cdot \frac{1}{4}$$

$$15 = 16R - 6$$

$$21 = 16R$$

$$16R = 21$$

$$R = \frac{21}{16}$$

Check:

L.H.S. $= \frac{5}{8}$

R.H.S. $= \frac{2}{3} \cdot \frac{21}{16} - \frac{1}{4} = \frac{7}{8} - \frac{2}{8} = \frac{5}{8}$

Hence, R $= \frac{21}{16}$ is the solution.

5. LCD = 20

$$20 \left(\frac{3x}{5} - \frac{2x + 3}{2} \right) = 20 \cdot \frac{1}{4}$$

$$20 \cdot \frac{3x}{5} - 20 \cdot \frac{2x + 3}{2} = 20 \cdot \frac{1}{4}$$

$$12x - 10(2x + 3) = 5$$

$$12x - 20x - 30 = 5$$

$$-8x = 35$$

$$x = -\frac{35}{8}$$

I   5.   (*continued*)

Check:

$$\text{L.H.S.} = \frac{3}{5}\left(-\frac{35}{8}\right) - \frac{2\left(-\frac{35}{8}\right)+3}{2} = -\frac{21}{8} - \frac{-\frac{23}{4}}{2} = -\frac{21}{8} + \frac{23}{8} = \frac{1}{4}$$

$$\text{R.H.S.} = \frac{1}{4}$$

Hence, x $= -\frac{35}{8}$ is the solution.

6. LCD $= 3m$

$$3M \cdot \frac{7}{3} = 3M\left(\frac{2}{M} - 5\right)$$

$$3M \cdot \frac{7}{3} = 3M \cdot \frac{2}{M} - 3M \cdot 5$$

$$7M = 6 - 15M$$
$$22M = 6$$
$$M = \frac{6}{22} \text{ or } \frac{3}{11}$$

Check:

$$\text{L.H.S.} = \frac{7}{3}$$

$$\text{R.H.S.} = \frac{2}{\frac{3}{11}} - 5 = \frac{22}{3} - \frac{15}{3} = \frac{7}{3}$$

Hence, M $= \frac{3}{11}$ is the solution.

7. LCD $= 3p$

$$3p\left(\frac{1}{p} - 2\right) = 3p \cdot \frac{5}{3}$$

$$3p \cdot \frac{1}{p} - 3p \cdot 2 = 3p \cdot \frac{5}{3}$$

$$3 - 6p = 5p$$
$$-11p = -3$$
$$p = \frac{3}{11}$$

Check:

$$\text{L.H.S.} = \frac{1}{\frac{3}{11}} - 2 = \frac{11}{3} - 2 = \frac{5}{3}$$

$$\text{R.H.S.} = \frac{5}{3}$$

Hence, p $= \frac{3}{11}$ is the solution.

8. LCD $= 4(3a - 4)$

$$4(3a - 4)\frac{5}{4} = 4(3a - 4)\frac{1}{3a - 4}$$

$$3a \cdot 5 - 4 \cdot 5 = 4$$
$$15a - 20 = 4$$
$$15a = 24$$
$$a = \frac{24}{15} \text{ or } \frac{8}{5}$$

Check:

$$\text{L.H.S.} = \frac{5}{4}$$

$$\text{R.H.S.} = \frac{1}{3 \cdot \frac{8}{5} - 4} = \frac{1}{\frac{24}{5} - \frac{20}{5}} = \frac{1}{\frac{4}{5}} = \frac{5}{4}$$

Hence, a $= \frac{8}{5}$ is the solution.

9. LCD $= d(d - 3)$

$$d(d - 3)\frac{3}{d - 3} = d(d - 3)\frac{2}{d}$$

$$3d = 2(d - 3)$$
$$3d = 2d - 6$$
$$d = -6$$

Check:

$$\text{L.H.S.} = \frac{3}{-6 - 3} = \frac{3}{-9} = -\frac{1}{3}$$

$$\text{R.H.S.} = \frac{2}{-6} = -\frac{1}{3}$$

Hence, d $= -6$ is the solution.

I   10.   LCD $= V - 1$

$$(V - 1)\frac{V + 1}{V - 1} = (V - 1)\left(2 + \frac{V}{V - 1}\right)$$

$$(V - 1)\frac{V + 1}{V - 1} = (V - 1)2 + (V - 1)\frac{V}{V - 1}$$

$$V + 1 = 2V - 2 + V$$

$$V + 1 = 3V - 2$$

$$-2V = -3$$

$$V = \frac{3}{2}$$

Check:

$$\text{L.H.S.} = \frac{\frac{3}{2} + 1}{\frac{3}{2} - 1} = \frac{\frac{5}{2}}{\frac{1}{2}} = \frac{5}{2} \cdot \frac{2}{1} = 5$$

$$\text{R.H.S.} = 2 + \frac{\frac{3}{2}}{\frac{3}{2} - 1} = 2 + \frac{\frac{3}{2}}{\frac{1}{2}} = 2 + 3 = 5$$

Hence, $V = \frac{3}{2}$ is the solution.

Answers to Assignments

1. $p = 7$ 3. $R = -\frac{3}{10}$ 5. $V = \frac{3}{4}$ 7. $m = \frac{3}{2}$ 9. $x = -\frac{11}{4}$

Unit 25

Solutions to Drill Exercises

I   1.   The ratio of foreign students to Canadian students

is $\frac{400}{4600}$ or $\frac{2}{23}$.

2.   The ratio of Canadian students to foreign students

is $\frac{4600}{400}$ or $\frac{23}{2}$.

3.   The ratio of foreign students to the total number of students

is $\frac{400}{5000}$ or $\frac{2}{25}$.

4.   The ratio of Canadian students to the total number of students

is $\frac{4600}{5000}$ or $\frac{23}{25}$.

II   5.   The ratio of width to length is $\dfrac{80 \text{ cm}}{1 \text{ m}}$.

Changing metres to centimetres, we have $\dfrac{80 \text{ cm}}{100 \text{ cm}}$.

Reducing $\dfrac{80 \text{ cm}}{100 \text{ cm}}$ to lowest form, we have $\dfrac{4}{5}$.

Hence, the ratio of width to length is 4 to 5.

6.   The density of the object is $\dfrac{12 \text{ g}}{5 \text{ cm}^3}$ or 2.4 g/cm^3.

III   7.   4:x :: 10:3 is the same as $\dfrac{4}{x} = \dfrac{10}{3}$

$$3x \cdot \dfrac{4}{x} = 3x \cdot \dfrac{10}{3}$$

$$12 = 10x$$

$$x = \dfrac{12}{10} \text{ or } 1.2$$

8.   2:5 :: 6:n is the same as $\dfrac{2}{5} = \dfrac{6}{n}$

$$5n \cdot \dfrac{2}{5} = 5n \cdot \dfrac{6}{n}$$

$$2n = 30$$

$$n = 15$$

IV   9.   Let n = the amount of chlorine in grams.

27 is to 48   as   10 is to n

$\dfrac{27}{48}$ reduces to $\dfrac{9}{16}$.

$$\dfrac{9}{16} = \dfrac{10}{n}$$

$$16n \cdot \dfrac{9}{16} = 16n \cdot \dfrac{10}{n}$$

$$9n = 160$$

$$n = \dfrac{160}{9}$$

$$= 17.\overline{7} \text{ or } 17.8 \text{ rounded to one decimal place.}$$

Check: $\dfrac{27}{48} = \dfrac{9}{16}$ and $\dfrac{10}{\dfrac{160}{9}} = 10 \cdot \dfrac{9}{160} = \dfrac{9}{16}$

Hence, 17.8 g of chlorine will combine with 10 g of calcium.

10.   Let n = the number of kilometres it can travel in 20 minutes.

3600 km is to 4 hr   as   n km is to 20 minutes

Changing hours to minutes, we have

3600 km is to 240 minutes   as   n km is to 20 minutes.

$\dfrac{3600}{240}$ reduces to 15

$$15 = \dfrac{n}{20}$$

$$20 \cdot 15 = 20 \cdot \dfrac{n}{20}$$

$$n = 300$$

Check: $\dfrac{3600 \text{ km}}{4 \text{ hr}} = \dfrac{3600 \text{ km}}{240 \text{ min}} = 15$ km/min and $\dfrac{300 \text{ km}}{20 \text{ min}} = 15$ km/min

Hence, it can travel 300 km in 20 minutes.

IV   11.   Let n = the amount partner A will receive.

78 000 − n = the amount partner B will receive.

3 is to 5   as   n is to 78 000 − n

$$\frac{3}{5} = \frac{n}{78\ 000 - n}$$

$$5(78\ 000 - n) \cdot \frac{3}{5} = 5(78\ 000 - n) \cdot \frac{n}{78\ 000 - n}$$

$$(78\ 000 - n)3 = 5n$$

$$234\ 000 - 3n = 5n$$

$$234\ 000 = 8n$$

$$n = 29\ 250$$

$$78\ 000 - 29\ 250 = 48\ 750$$

Check: $\dfrac{28\ 250}{48\ 750} = \dfrac{3 \times 9750}{5 \times 9750} = \dfrac{3}{5}$

Hence, one partner will receive \$29 250 and the other will receive \$48 750.

12.   From the equality, 1 kg = 1000 g, we select

$\dfrac{1000\ g}{1\ kg}$ as the conversion ratio.

2.4 kg x $\dfrac{1000\ g}{1\ kg}$ = 2400 g

Hence, 2.4 kg = 2400 g .

13.   From the equality, 1 mile = 1.609 km, we select

$\dfrac{1.609\ km}{1\ mi}$ as the conversion ratio.

125 mi · $\dfrac{1.609\ km}{1\ mi}$ = 201.125

Hence, 125 mi = 201 km
(rounded to the nearest kilometre)

14.   From the equalities, 1 hour = 60 minutes and 1 minute = 60 seconds, we select

$\dfrac{1\ min}{60\ s}$ and $\dfrac{1\ hr}{60\ min}$ as the conversion ratios.

4050 s · $\dfrac{1\ min}{60\ s}$ · $\dfrac{1\ hr}{60\ min}$

= $\dfrac{4050\ hr}{60 \times 60}$

=   1.125 hr

Hence, 4050 s = 1 hr (rounded to the nearest hour)

IV   15.   From 1 km = 1000 m, we select the ratio $\dfrac{1000 \text{ m}}{1 \text{ km}}$.

From 1 hour = 60 minutes, we select the ratio $\dfrac{1 \text{ hr}}{60 \text{ min}}$.

From 1 min = 60 seconds, we select the ratio $\dfrac{1 \text{ min}}{60 \text{ s}}$.

$$\frac{45 \text{ km}}{\text{hr}} = \frac{45 \text{ km}}{\text{hr}} \times \frac{1000 \text{ m}}{1 \text{ km}} \times \frac{1 \text{ hr}}{60 \text{ min}} \times \frac{1 \text{ min}}{60 \text{ s}}$$

$$= \frac{45 \times 1000 \text{ m}}{60 \times 60 \text{ s}}$$

$$= 12.5 \text{ m/s}$$

Hence, 45 km/m = 12.5 m/s.

16.   From 1 gal = 4 qt, we select the ratio $\dfrac{4 \text{ qt}}{1 \text{ gal}}$.

From 1 qt = 1.14 L, we select the ratio $\dfrac{1.14 \text{ L}}{1 \text{ qt}}$.

From 1 L = 1000 mL, we select the ratio $\dfrac{1000 \text{ mL}}{1 \text{ L}}$.

$$2.4 \text{ gal} = 2.4 \text{ gal} \times \frac{4 \text{ qt}}{1 \text{ gal}} \times \frac{1.14 \text{ L}}{1 \text{ qt}} \times \frac{1000 \text{ mL}}{1 \text{ L}}$$

$$= 2.4 \times 4 \times 1.14 \times 1000 \text{ mL}$$

$$= 10\,944 \text{ mL}$$

Hence, 2.4 gal = 10 944 mL.

17.   From 1 in = 2.54 cm, we select the ratio $\dfrac{2.54 \text{ cm}}{1 \text{ in}}$.

From 1m = 100 cm, we select the ratio $\dfrac{1 \text{ m}}{100 \text{ cm}}$.

$$65 \text{ in} = 65 \text{ in} \times \frac{2.54 \text{ cm}}{1 \text{ in}} \times \frac{1 \text{ m}}{100 \text{ cm}}$$

$$= \frac{65 \times 2.54 \text{ m}}{100}$$

$$= 1.651 \text{ m}$$

Hence, 65 in = 1.651 m.

I   18.

$$1 \text{ yd} = 3 \text{ ft}$$

$$1.5 \text{ yd} = 1.5 \text{ yd} \times \frac{3 \text{ ft}}{1 \text{ yd}}$$

$$= 4.5 \text{ ft}$$

19.

$$1.14 \text{ L} = 1 \text{ qt}$$

$$5\text{L} = 5 \text{ L} \times \frac{1 \text{ qt}}{1.14 \text{ L}}$$

$$= 4.4 \text{ qt}$$

I   20.   64.8 mg = 1 grain

350 mg = 350 mg x $\dfrac{1 \text{ grain}}{64.8 \text{ mg}}$

= 5.4 grains

21.   28.4 mL = 1 fl oz (Can)

1 fl oz (Can) = 0.96 fl oz (U.S.)

62 mL = 62 mL x $\dfrac{1 \text{ fl oz (Can)}}{28.4 \text{ mL}}$

x $\dfrac{0.96 \text{ fl oz (U.S.)}}{1 \text{ fl oz (Can)}}$

= 2.1 fl oz (U.S.)

22.   1.609 km = 1 mi

240 km = 240 km x $\dfrac{1 \text{ mi}}{1.609 \text{ km}}$

= 149.2 mi

23.   1 lb = 454 g

12.5 lb = 12.5 lb x $\dfrac{454 \text{ g}}{1 \text{ lb}}$

= 5675.0 g

24.   0.4047 ha = 1 acre

10 ha = 10 ha x $\dfrac{1 \text{ acre}}{0.4047 \text{ ha}}$

= 24.7 acres

25.   1 fl oz = 28.4 mL

85 fl oz = 85 fl oz x $\dfrac{28.4 \text{ mL}}{1 \text{ fl oz}}$

= 2414.0 mL

26.   12 in = 1 ft

49 in = 49 in x $\dfrac{1 \text{ ft}}{12 \text{ in}}$

= 4.1 ft

27. 1.20 gal (U.S.) = 1 gal (Can)

60 gal (U.S.) = 60 gal (U.S.)

x $\dfrac{1 \text{ gal (Can)}}{1.20 \text{ gal (U.S.)}}$

= 50.0 gal (Can)

28.   16 oz = 1 lb

31.5 oz = 31.5 oz x $\dfrac{1 \text{ lb}}{16 \text{ oz}}$

= 2.0 lb

29.   1 mi = 1.609 km

121 mi = 121 mi x $\dfrac{1.609 \text{ km}}{1 \text{ mi}}$

= 194.7 km

30.   1760 yd = 1 mi

2460 yd = 2460 yd x $\dfrac{1 \text{ mi}}{1760 \text{ yd}}$

= 1.4 mi

31.   1 yd = 91.44 cm

0.5 yd = 0.5 yd x $\dfrac{91.44 \text{ cm}}{1 \text{ yd}}$

= 45.7 cm

32.   1 ft = 30.48 cm

100 cm = 1 m

8 ft = 8 ft x $\dfrac{30.48 \text{ cm}}{1 \text{ ft}}$ x $\dfrac{1 \text{ m}}{100 \text{ cm}}$

= 2.4 m

I   33.   1 gal (Can)   = 4 qt

1 qt   = 1.14 L

$$29 \text{ gal (Can)} = 29 \text{ gal (Can)} \times \frac{4 \text{ qt}}{1 \text{ gal (Can)}} \times \frac{1.14 \text{ L}}{1 \text{ qt}}$$

= 132.2 L

34.   2.54 cm   = 1 in

$$3.6 \text{ cm} = 3.6 \text{ cm} \times \frac{1 \text{ in}}{2.54 \text{ cm}}$$

= 1.4 in

35.   16 oz   = 1 lb

1 lb   = 454 g

1000 g   = 1 kg

$$124 \text{ oz} = 124 \text{ oz} \times \frac{1 \text{ lb}}{16 \text{ oz}} \times \frac{454 \text{ g}}{1 \text{ lb}} \times \frac{1 \text{ kg}}{1000 \text{ g}}$$

= 3.5 kg

36.   1 g   = 1000 mg

64.8 mg   = 1 grain

$$0.8 \text{ g} = 0.8 \text{ g} \times \frac{1000 \text{ mg}}{1 \text{ g}} \times \frac{1 \text{ grain}}{64.8 \text{ mg}}$$

= 12.3 grains

37.   0.96 fl oz (U.S.)   = 1 fl oz (Can)

20 fl oz   = 1 pt

2 pt   = 1 qt

$$40 \text{ fl oz (U.S.)} = 40 \text{ fl oz (U.S.)} \times \frac{1 \text{ fl oz (Can)}}{0.96 \text{ fl oz (U.S.)}} \times \frac{1 \text{ pt}}{20 \text{ fl oz}} \times \frac{1 \text{ qt}}{2 \text{ pt}}$$

= 1.0 qt (Can)

38.   1 kg   = 1000 g

454 g   = 1 lb

1 lb   = 16 oz

$$3.5 \text{ kg} = 3.5 \text{ kg} \times \frac{1000 \text{ g}}{1 \text{ kg}} \times \frac{1 \text{ lb}}{454 \text{ g}} \times \frac{16 \text{ oz}}{1 \text{ lb}}$$

= 123.3 oz

39.   1 gal   = 4 qt

1 qt   = 1.14 L

1 L   = 1000 mL

$$0.75 \text{ gal (Can)} = 0.75 \text{ gal (Can)} \times \frac{4 \text{ qt}}{1 \text{ gal}} \times \frac{1.14 \text{ L}}{1 \text{ qt}} \times \frac{1000 \text{ mL}}{1 \text{ L}}$$

= 3420 mL

I   40.

$$1 \text{ lb} = 454 \text{ g}$$

$$1000 \text{ g} = 1 \text{ kg}$$

$$265 \text{ lb} = 265 \cancel{\text{lb}} \times \frac{454 \cancel{\text{g}}}{1 \cancel{\text{lb}}} \times \frac{1 \text{ kg}}{1000 \cancel{\text{g}}}$$

$$= 120.3 \text{ kg}$$

41.

$$28.4 \text{ mL} = 1 \text{ fl oz (Can)}$$

$$1 \text{ fl oz (Can)} = 0.96 \text{ fl oz (U.S.)}$$

$$782 \text{ mL} = 782 \cancel{\text{mL}} \times \frac{1 \cancel{\text{fl oz (Can)}}}{28.4 \cancel{\text{mL}}} \times \frac{0.96 \text{ fl oz (U.S.)}}{1 \cancel{\text{fl oz (Can)}}}$$

$$= 26.4 \text{ fl oz (U.S.)}$$

II   42.

$$16 \text{ C}° = 16 \cancel{\text{C}°} \times \frac{9 \text{ F}°}{5 \cancel{\text{C}°}}$$

$$= 28.8 \text{ F}°$$

43.

$$1.8 \text{ C}° = 1.8 \cancel{\text{C}°} \times \frac{9 \text{ F}°}{5 \cancel{\text{C}°}}$$

$$= 3.2 \text{ F}°$$

44.

$$25 \text{ F}° = 25 \cancel{\text{F}°} \times \frac{5 \text{ C}°}{9 \cancel{\text{F}°}}$$

$$= 13.9 \text{ C}°$$

45.

$$9.6 \text{ F}° = 9.6 \cancel{\text{F}°} \times \frac{5 \text{ C}°}{9 \cancel{\text{F}°}}$$

$$= 5.3 \text{ C}°$$

III   46.

$$12° \text{ C} = \frac{9}{5}(12) + 32$$

$$= 21.6 + 32$$

$$= 53.6° \text{ F}$$

47.

$$-5° \text{ C} = \frac{9}{5}(-5) + 32$$

$$= -9 + 32$$

$$= 23° \text{ F}$$

48.

$$75° \text{ C} = \frac{9}{5}(75) + 32$$

$$= 135 + 32$$

$$= 167° \text{ F}$$

49.

$$92° \text{ C} = \frac{9}{5}(92) + 32$$

$$= 165.6 + 32$$

$$= 197.6° \text{ F}$$

50.

$$0° \text{ C} = \frac{9}{5}(0) + 32$$

$$= 0 + 32$$

$$= 32° \text{ F}$$

51.

$$3.2° \text{ C} = \frac{9}{5}(3.2) + 32$$

$$= 5.8 + 32$$

$$= 37.8° \text{ F}$$

52.

$$-21.5° \text{ C} = \frac{9}{5}(-21.5) + 32$$

$$= -38.7 + 32$$

$$= 6.7° \text{ F}$$

53.

$$-1.7° \text{ C} = \frac{9}{5}(-1.7) + 32$$

$$= -3.1 + 32$$

$$= 28.9° \text{ F}$$

54.

$$30° \text{ C} = \frac{9}{5}(30) + 32$$

$$= 54 + 32$$

$$= 86° \text{ F}$$

55.

$$21.8° \text{ C} = \frac{9}{5}(21.8) + 32$$

$$= 39.2 + 32$$

$$= 71.2° \text{ F}$$

III   56. $-12.4°\text{ C} = \frac{9}{5}(-12.4) + 32$

$= -22.3 + 32$

$= 9.7°\text{ F}$

57. $-30°\text{ C} = \frac{9}{5}(-30) + 32$

$= -54 + 32$

$= -22°\text{ F}$

58. $100°\text{ C} = \frac{9}{5}(100) + 32$

$= 180 + 32$

$= 212°\text{ F}$

59. $-42.6°\text{ C} = \frac{9}{5}(-42.6) + 32$

$= -76.7 + 32$

$= -44.7°\text{ F}$

60. $51.3°\text{ C} = \frac{9}{5}(51.3) + 32$

$= 92.3 + 32$

$= 124.3°\text{ F}$

61. $37°\text{ C} = \frac{9}{5}(37) + 32$

$= 66.6 + 32$

$= 98.6°\text{ F}$

IV   62. $80°\text{ F} = \frac{5}{9}(80 - 32)$

$= \frac{5}{9}(48)$

$= 26.7°\text{ C}$

63. $32°\text{ F} = \frac{5}{9}(32 - 32)$

$= \frac{5}{9}(0)$

$= 0°\text{ C}$

64. $59°\text{ F} = \frac{5}{9}(59 - 32)$

$= \frac{5}{9}(27)$

$= 15°\text{ C}$

65. $98.6°\text{ F} = \frac{5}{9}(98.6 - 32)$

$= \frac{5}{9}(66.6)$

$= 37°\text{ C}$

66. $21°\text{ F} = \frac{5}{9}(21 - 32)$

$= \frac{5}{9}(-11)$

$= -6.1°\text{ C}$

67. $0°\text{ F} = \frac{5}{9}(-32)$

$= \frac{5}{9}(-32)$

$= -17.8°\text{ C}$

68. $15°\text{ F} = \frac{5}{9}(15 - 32)$

$= \frac{5}{9}(-17)$

$= -9.4°\text{ C}$

69. $39.5°\text{ F} = \frac{5}{9}(39.5 - 32)$

$= \frac{5}{9}(7.5)$

$= 4.2°\text{ C}$

70. $-10°\text{ F} = \frac{5}{9}(-10 - 32)$

$= \frac{5}{9}(-42)$

$= -23.3°\text{ C}$

71. $64.5°\text{ F} = \frac{5}{9}(64.5 - 32)$

$= \frac{5}{9}(32.5)$

$= 18.1°\text{ C}$

72. $184.7°\text{ F} = \frac{5}{9}(184.7 - 32)$

$= \frac{5}{9}(152.7)$

$= 84.8°\text{ C}$

73. $212°\text{ F} = \frac{5}{9}(212 - 32)$

$= \frac{5}{9}(180)$

$= 100°\text{ C}$

74. $124.6° \text{ F} = \frac{5}{9}(124.6 - 32)$

$= \frac{5}{9}(92.6)$

$= 51.4° \text{ C}$

75. $70° \text{ F} = \frac{5}{9}(70 - 32)$

$= \frac{5}{9}(38)$

$= 21.1° \text{ C}$

76. $-5° \text{ F} = \frac{5}{9}(-5 - 32)$

$= \frac{5}{9}(-37)$

$= -20.6° \text{ C}$

77. $8.7° \text{ F} = \frac{5}{9}(8.7 - 32)$

$= \frac{5}{9}(-23.3)$

$= -12.9° \text{ C}$

Answers to Assignments

I   1. $\frac{3}{17}$ 3. $\frac{3}{20}$

II   5. $\frac{3}{20}$

III   7. $n = \frac{24}{5}$

IV   9.   144 km   11.   208 kg (rounded to the nearest kg)

V   13. 0.125 L

VI   15. 98 in   17.   114.5 oz   19.   972.0 mg   21.   553.8 mL

        23. 10.9 ha   25.   99.8 km   27.   1320.0 yd   29.   14.1 lb

        31. 6.9 ft   33.   1.6 g   35.   0.8 qt   37.   1892.0 mL

        39. 67.6 kg   41.   21.6°F   43.   17.8°C   45.   42.8°F

        47. 176°F   49.   65.7°F   51.   −25.6°F   53.   14.9°F

        55. 76.6°F   57.   35.4°F   59.   180.5°F   61.   17.8°C

        63. 34.4°C   65.   87.8°C   67.   5.8°C   69.   28.1°C

        71. −30.6°C   73.   98.9°C   75.   24.7°C

Unit 26

Solutions to Drill Exercises

I   1.   Let x = number.

     x + 6 = the number plus 6.

     $\frac{x + 6}{5}$ = the sum (of the number plus 6) divided by 5.

     "The sum (of the number plus 6) divided by 5 is 3."

$$\frac{x + 6}{5} = 3$$

$$5\left(\frac{x + 6}{5}\right) = 5 \cdot 3$$

$$x + 6 = 15$$

$$x = 9$$

If x = 9, $\frac{x + 6}{5} = \frac{9 + 6}{5} = 3.$

Hence, the number is 9.

I   2.   Let x = the number.

$\dfrac{12}{100} x = 12\%$ of the number.

$x + \dfrac{12}{100} x$ = the number increased by 12%.

"A number increased by 12% is 84."

$$x + \dfrac{12}{100} x = 84$$

$$100\left(x + \dfrac{12}{100} x\right) = 100 \cdot 84$$

$$100x + 100\,\dfrac{12}{100}\, x = 100 \cdot 84$$

$$100x + 12x = 8400$$

$$112x = 8400$$

$$x = \dfrac{8400}{112}$$

$$x = 75$$

If $x = 75$, then $75 + \dfrac{12}{100} \cdot 75 = 75 + 9 = 84$.

Hence, the number is 75.

3.   Let x = the larger number.

$\dfrac{1}{5} x - 12$ = the smaller number.

"The sum of the two numbers is 78."

$$x + \dfrac{1}{5} x - 12 = 78$$

$$5\left(x + \dfrac{1}{5} x - 12\right) = 5 \cdot 78$$

$$5x + 5 \cdot \dfrac{1}{5} x - 5 \cdot 12 = 5 \cdot 78$$

$$5x + x - 60 = 390$$

$$6x = 450$$

$$x = \dfrac{450}{6} \text{ or } 75$$

If $x = 75$, $\dfrac{1}{5} x - 12 = \dfrac{1}{5}(75) - 12 = 15 - 12 = 3$.

$75 + 3 = 78$ is the correct sum.

Hence, the numbers are 3 and 75.

4.   Let x = the amount of interest he has to pay.

"13% of \$3000 = x"

$$\dfrac{13}{100} \cdot 3000 = x$$

$$x = \dfrac{13}{100} \cdot 3000$$

$$100x = \overset{1}{\cancel{100}} \cdot \dfrac{13}{\underset{1}{\cancel{100}}} \cdot 3000$$

$$100x = 39\,000$$

$$x = 390$$

Hence, he has to pay \$390 interest.

I   5.   Let x = the amount he was making before he got the new contract.

$$x + \frac{15x}{100} = \text{his present pay}$$

$$x + \frac{15x}{100} = 25\ 300$$

$$100\left(x + \frac{15x}{100}\right) = 100 \cdot 25\ 300$$

$$100x + \overset{1}{\cancel{100}} \cdot \frac{15x}{\underset{1}{\cancel{100}}} = 100 \cdot 25\ 300$$

$$100x + 15x = 2\ 530\ 000$$

$$115x = 2\ 530\ 000$$

$$x = 22\ 000$$

If x = 22 000, $22\ 000 + \frac{15}{100} \cdot 22\ 000 = 22\ 000 + 3300 = 25\ 300$.

Hence, he was making $22 000 before he got the new contract.

6.   (a)   Let x = the profit (in percent).

$$18 + 18\frac{x}{100} = 24$$

$$100\left(18 + 18\frac{x}{100}\right) = 100 \cdot 24$$

$$100 \cdot 18 + \overset{1}{\cancel{100}} \cdot \frac{18x}{\underset{1}{\cancel{100}}} = 100 \cdot 24$$

$$1800 + 18x = 2400$$

$$18x = 600$$

$$x = \frac{100}{3}$$

If $x = \frac{100}{3}$, $18 + 18\frac{\frac{100}{3}}{100} = 18 + 18 \cdot \frac{1}{3} = 18 + 6 = 24$.

Hence, the profit is $\frac{100}{3}\% = 33\frac{1}{3}\%$ or   33.33%.

I   6.   *(continued)*

    (b)   Let y = the profit (in percent).

$$18 + 18\frac{x}{100} = 21$$

$$100\left(18 + 18\frac{x}{100}\right) = 100 \cdot 21$$

$$100 \cdot 18 + \overset{1}{\cancel{100}} \cdot \frac{18x}{\underset{1}{\cancel{100}}} = 2100$$

$$1800 + 18x = 2100$$

$$18x = 300$$

$$x = \frac{50}{3}$$

If $x = \frac{50}{3}$, $\quad 18 + 18\left(\dfrac{\frac{50}{3}}{100}\right) = 18 + \frac{900}{300} = 18 + 3 = 21$.

Hence, the profit is $\dfrac{50}{3}\% = 16\frac{2}{3}\%$ or 16.67%.

7.   Let x = the amount of interest he paid that year.

$$15\% \text{ of } \$15\,000 = x$$

$$\frac{15}{100} \times 15\,000 = x$$

$$x = \frac{15}{100} \cdot 15\,000$$

$$x = 2250$$

Let y = the percent of the \$5000 that was interest

$$\frac{y}{100} \cdot 5000 = 2250$$

$$50y = 2250$$

$$y = 45$$

If $y = 45$, $\dfrac{45}{100} \cdot 5000 = 2250$.

Hence, 45% of this payment is interest.

I   8.   Let p = the amount invested at 18%.

60 000 – p = the amount invested at 12%.

$$\frac{18}{100}p + \frac{12}{100}(60\,000 - p) = 9300$$

$$\frac{1}{\cancel{100}} \cdot \frac{18p}{\cancel{100}} + \frac{1}{\cancel{100}} \cdot \frac{12}{\cancel{100}}(60\,000 - p) = 100 \cdot 9300$$

$$18p + 12(60\,000 - p) = 930\,000$$

$$18p + 720\,000 - 12p = 930\,000$$

$$6p = 210\,000$$

$$p = 35\,000$$

If p = 35 000, then 60 000 – p = 25 000.

Hence, $35 000 has been invested at 18% and $25 000 has been invested at 12%.

Check:

Interest earned at 18% is $\frac{18}{100}(35\,000) = 6300$.

Interest earned at 12% is $\frac{12}{100}(25\,000) = 3000$.

Total interest earned is 9300. That is correct.

9.   Let L = the length of the rectangle.

$\frac{2}{3}L + 2 =$ the width of the rectangle.

"The perimeter is 94 cm."   (P = 2L + 2W)

$$2L + 2\left(\frac{2}{3}L + 2\right) = 94$$

$$2L + \frac{4}{3}L + 4 = 94$$

$$3\left(2L + \frac{4}{3}L + 4\right) = 3 \cdot 94$$

$$6L + 4L + 12 = 282$$

$$10L = 270$$

$$L = 27$$

If L = 27, then $\frac{2}{3}L + 2 = \frac{2}{3}(27) + 2 = 18 + 2 = 20$

Hence, the length is 27 cm and the width is 20 cm.

Check: 2(27) + 2(20) = 54 + 40 = 94

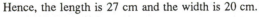

I   10.   Let S = the length of the first side.

$\frac{2}{5}$ S + 4 = the length of the second side.

$\frac{8}{11}$ (S + $\frac{2}{5}$ S + 4) = the length of the third side.

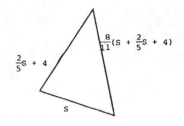

"The perimeter of the triangle is 19 m."

$$S + \frac{2}{5}S + 4 + \frac{8}{11}(S + \frac{2}{5}S + 4) = 19$$

$$S + \frac{2}{5}S + 4 + \frac{8}{11}S + \frac{16}{55}S + \frac{32}{11} = 19$$

(LCD = 55)

$$55\left(S + \frac{2}{5}S + 4 + \frac{8}{11}S + \frac{16}{55}S + \frac{32}{11}\right) = 55\,(19)$$

$$55S + 22S + 220 + 40S + 16S + 160 = 1045$$

$$133S = 665$$

$$S = 5$$

If the length of the first side is 5, the length of the second side is $\frac{2}{5}$S + 4 = $\frac{2}{5}$ (5) + 4 = 6, and the length of the third side is $\frac{8}{11}$ (S + $\frac{2}{5}$ S + 4) = $\frac{8}{11}$ (5 + 2 + 4) = $\frac{8}{11}$ (11) = 8

Hence, the lengths of the three sides are 5m, 6m, and 8m.

Check: 5 + 6 + 8 = 19

11.   Step 1.   Let W = the width of the original rectangle.

$\frac{5}{4}$ W – 4 = the length of the original rectangle.

"The perimeter is 82 cm."   (P = 2L + 2W)

$$2\left(\frac{5}{4}W - 4\right) + 2W = 82$$

$$\frac{10W}{4} - 8 + 2W = 82$$

$$\frac{5W}{2} - 8 + 2W = 82$$

$$5W - 16 + 4W = 164$$

$$9W = 180$$

$$W = \frac{180}{9} = 20$$

If W = 20, then $\frac{5}{4}$ W – 4 = $\frac{5}{4}$ (20) – 4 = 25 – 4 = 21

Hence, the length is 21 cm and the width is 20 cm for the original rectangle.

Check: 2(21) + 2(20) = 42 + 40 = 82

I   11.   *(continued)*

Step 2.   Let L_1 = the length of the original rectangle.

L_2 = the length of the new rectangle.

$$L_2 = L_1 + \frac{2}{5}W = 21 + \frac{2}{5}(20) = 21 + 8 = 29$$

For the new rectangle, the length is 29 cm and the width is 20 cm.

Hence, the perimeter of the new rectangle

$$= 2(29) + 2(20)$$

$$= 58 + 40$$

$$= 98 \text{ cm.}$$

12.   Let x   = the length of one piece.

$\frac{4}{7}x - \frac{1}{2}$ = the length of the other piece.

"A beam 5 m long is cut into two pieces."

$$x + \frac{4}{7}x - \frac{1}{2} = 5$$

$$14\left(x + \frac{4}{7}x - \frac{1}{2}\right) = 14 \cdot 5$$

$$14x + 8x - 7 = 70$$

$$22x = 77$$

$$x = \frac{77}{22}$$

$$x = \frac{7}{2}$$

If the length of the first piece is $\frac{7}{2}$ m, then the length of the second piece is

$$\frac{4}{7}x - \frac{1}{2} = \frac{4}{7} \cdot \frac{7}{2} - \frac{1}{2} = \frac{4}{2} - \frac{1}{2} = \frac{3}{2} \text{ m.}$$

Hence, the lengths of the two pieces are

$$\frac{7}{2} = 3.5 \text{ m} \quad \text{and} \quad \frac{3}{2} = 1.5 \text{ m.}$$

Check: $\frac{7}{2} + \frac{3}{2} = 5$

I   13.   Let x   =   the number of mℓ of water to be added

500 + x   =   the number of mℓ in a new 6% solution.

$$\boxed{\begin{array}{l}\text{Pure acid}\\\text{in 15\%}\\\text{solution}\end{array}} \quad = \quad \boxed{\begin{array}{l}\text{Pure acid}\\\text{in 6\%}\\\text{solution}\end{array}}$$

15% of 500 mℓ   =   6% of (500 + x) mℓ

$$\frac{15}{100}(500) \quad = \quad \frac{6}{100}(500 + x)$$

Solve the equation.

$$\frac{15}{100}(500) \quad = \quad \frac{6}{100}(500 + x)$$

$$7500 \quad = \quad 3000 + 6x$$

$$6x \quad = \quad 4500$$

$$x \quad = \quad 750$$

Hence, we have to add 750 mℓ.

Check: $15\% \text{ of } 500 = \frac{15}{100} \times 500 = 75 \text{ mℓ}$

$6\% \text{ of } 1250 = \frac{6}{100} \times 1250 = 75 \text{ mℓ}$

14.   Let x   =   the number of kg of the 65% alloy required.

150 000 − x   =   the number of kg of the 20% alloy required.

$$\boxed{\begin{array}{l}\text{Pure nickel in}\\\text{the 65\% alloy}\end{array}} \quad + \quad \boxed{\begin{array}{l}\text{Pure nickel in}\\\text{the 20\% alloy}\end{array}} \quad = \quad \boxed{\begin{array}{l}\text{Pure nickel in}\\\text{the 35\% alloy}\end{array}}$$

65% of x kg   +   20% of (150 000 − x) kg   =   35% of 150 000 kg

$$\frac{65}{100}x \quad + \quad \frac{20}{100}(150\,000 - x) \quad = \quad \frac{35}{100}(150\,000)$$

Solve the equation.

$$\frac{65}{100}x + \frac{20}{100}(150\,000 - x) \quad = \quad \frac{35}{100}(150\,000)$$

$$65x + 20(150\,000 - x) \quad = \quad 35(150\,000)$$

$$65x + 3\,000\,000 - 20x \quad = \quad 5\,250\,000$$

$$45x \quad = \quad 2\,250\,000$$

$$x \quad = \quad 50\,000$$

If x = 50 000, then 150 000 − x = 100 000.

Hence, to obtain 150 000 kg of alloy containing 35% nickel, we require 50 000 kg of the 65% alloy and 100 000 kg of the 20% alloy.

Check: 50 000 + 100 000 = 150 000

I   15.   Let x = the number of kg of copper to be added.

40 + x = the number of kg of the new alloy.

Pure zinc in original alloy		Pure zinc in new alloy

$$= $$

8 kg   =   10% of (40 + x)

$$8 \quad = \quad \frac{10}{100}(40 + x)$$

Solve the equation.

$$8 \quad = \quad \frac{10}{100}(40 + x)$$

$$800 \quad = \quad 400 + 10x$$

$$10x \quad = \quad 400$$

$$x \quad = \quad 40$$

Hence, 40 kg of copper must be added.

Check: $\frac{10}{100}$ x 80 = 8 kg

Answers to Assignments

I   1.   35   3.   364   5.   48   7.   20% of cost price

9.   16%   11. $4 200.00   13. $3000.00 at 17%, $2000.00 at 15%

15. length = 90 cm, width = 60 cm   17. 12 cm, 5 cm, 14 cm   19. 16 cm, 39 cm

21. $3\frac{1}{3}$ L   23. 35 L   25. 390 kg of 90% alloy, 520 kg of 20% alloy

Unit 27

Solutions to Drill Exercises

I   1. $G = \frac{K - H}{K}$ 2. $F = C - \frac{AB}{X}$

$$= \frac{10 - 8}{10} \qquad\qquad\qquad = 10 - \frac{100(0)}{1000}$$

$$= \frac{2}{10} \qquad\qquad\qquad\qquad = 10 - \frac{0}{1000}$$

$$\therefore G = \frac{1}{5} \qquad\qquad\qquad = 10 - 0$$

$$\therefore F = 10$$

I   3. $V_2 = \left(\dfrac{2m_1}{m_1 + m_2}\right) V_1$

   $= \left(\dfrac{2(20)}{20 + 10}\right) 27$

   $= \dfrac{40}{30} \cdot 27$

   $\therefore V_2 = 36$

4. $S = P(1 + rt)$

   $= 1000\left[1 + (.16)\left(\dfrac{60}{360}\right)\right]$

   $= 1000[1 + .0266667]$

   $= 1000[1.0266667]$

   $= 1026.6667$

   $\therefore S = 1026.67$

5. $C = \dfrac{5}{9}(F - 32)$

   $= \dfrac{5}{9}(20 - 32)$

   $= \dfrac{5}{9}(-12)$

   $= -6.66667$

   $\therefore C = -6.7°$

6. $d = \dfrac{i}{1 + in}$

   $= \dfrac{0.015}{1 + (.015)(24)}$

   $= \dfrac{0.015}{1.36}$

   $= .01102941$

   $\therefore d = .011$

II   7. $y - 2x = 7$

   $-2x = -y + 7$

   $\dfrac{-2x}{-2} = \dfrac{-y + 7}{-2}$

   $x = \dfrac{(-y + 7)(-1)}{-2(-1)}$

   $x = \dfrac{y - 7}{2}$

   Therefore, $x = \dfrac{y - 7}{2}$

8. $15x = 2y - 26$

   $-2y = -15x - 26$

   $\dfrac{-2y}{-2} = \dfrac{-15x - 26}{-2}$

   $y = \dfrac{(-15x - 26)(-1)}{-2(-1)}$

   $= \dfrac{15x + 26}{2}$

   Therefore, $y = \dfrac{15x + 26}{2}$

9. $\dfrac{1}{2}x + \dfrac{13}{6}y = 4$

   $6\left(\dfrac{1}{2}x + \dfrac{13}{6}y\right) = 6 \cdot 4$

   $6 \cdot \dfrac{1}{2}x + 6 \cdot \dfrac{13}{6}y = 6 \cdot 4$

   $3x + 13y = 24$

   $13y = -3x + 24$

   $\dfrac{13y}{13} = \dfrac{-3x + 24}{13}$

   $y = \dfrac{-3x + 24}{13}$

   Therefore, $y = \dfrac{-3x + 24}{13}$.

10. $y = \dfrac{3}{4}x - \dfrac{2}{3}$

   $12 \cdot y = 12\left(\dfrac{3}{4}x - \dfrac{2}{3}\right)$

   $12y = 9x - 8$

   $-9x = -12y - 8$

   $\dfrac{\overset{1}{\cancel{-9x}}}{\underset{1}{\cancel{-9}}} = \dfrac{-12y - 8}{-9}$

   $x = \dfrac{(-12y - 8)(-1)}{-9(-1)}$

   Therefore, $x = \dfrac{12y + 8}{9}$.

II   11. $\dfrac{5}{2}x = -4y + 1$

$5x = -8y + 2$

$8y = -5x + 2$

$y = \dfrac{-5x + 2}{8}$

Therefore, $y = \dfrac{-5x + 2}{8}$

12. $C = 2\pi r$

$\dfrac{C}{2\pi} = \dfrac{\overset{1}{\cancel{2\pi r}}}{\underset{1}{\cancel{2\pi}}}$

Therefore, $r = \dfrac{C}{2\pi}$

13. $\dfrac{V_1}{V_2} = \dfrac{P_1}{P_2}$

$\overset{1}{\cancel{V_2}}\, P_2\, \dfrac{V_1}{\underset{1}{\cancel{V_2}}} = V_2\, \overset{1}{\cancel{P_2}}\, \dfrac{P_1}{\underset{1}{\cancel{P_2}}}$

$P_2 V_1 = V_2 P_1$

$\dfrac{P_2\, \overset{1}{\cancel{V_1}}}{\underset{1}{\cancel{V_1}}} = \dfrac{V_2 P_1}{V_1}$

Therefore, $P_2 = \dfrac{V_2 P_1}{V_1}$.

14. $S = \dfrac{QV}{Ib}$

$IbS = \overset{1}{\cancel{Ib}}\, \dfrac{QV}{\underset{1}{\cancel{Ib}}}$

$IbS = QV$

$\dfrac{IbS}{Q} = \dfrac{\overset{1}{\cancel{Q}}V}{\underset{1}{\cancel{Q}}}$

Therefore, $V = \dfrac{IbS}{Q}$

15. $A = P(1 + i)^n$

$\dfrac{A}{(1 + i)^n} = \dfrac{P\,\overset{1}{\cancel{(1 + i)^n}}}{\underset{1}{\cancel{(1 + i)^n}}}$

Therefore, $P = \dfrac{A}{(1 + i)^n}$

16. $T = \dfrac{M(g + a)}{2}$

$2T = M(g + a)$

$\dfrac{2T}{g + a} = \dfrac{M\,\overset{1}{\cancel{(g + a)}}}{\underset{1}{\cancel{g + a}}}$

Therefore, $M = \dfrac{2T}{g + a}$

17. $F = \dfrac{Gm_1 m_2}{r_2}$

$r_2 F = \overset{1}{\cancel{r_2}} \cdot \dfrac{Gm_1 m_2}{\underset{1}{\cancel{r_2}}}$

$r_2 F = Gm_1 m_2$

$\dfrac{r_2 F}{Gm_2} = \dfrac{\overset{1}{\cancel{G}}m_1 \overset{1}{\cancel{m_2}}}{\underset{1\ \ 1}{\cancel{Gm_2}}}$

Therefore, $m_1 = \dfrac{r_2 F}{Gm_2}$

18. $N = L - C$

$C = L - N$

II   19. $D = P - FA$

$FA = P - D$

$$\frac{\cancel{F}^{1} A}{\cancel{F}_{1}} = \frac{P - D}{F}$$

Therefore, $A = \dfrac{P - D}{F}$.

20. $P = S_1 + S_2 + S_3$

$-S_2 = S_1 + S_3 - P$

$(-1)(-S_2) = (-1)(S_1 + S_3 - P)$

$S_2 = P - S_1 - S_3$

Therefore, $S_2 = P - S_1 - S_3$.

21. $y = mx + b$

$-mx = b - y$

$$\frac{\cancel{-m}^{1}\cancel{x}}{\cancel{-x}_{1}} = \frac{b - y}{-x}$$

$m = \dfrac{(b - y)(-1)}{-x(-1)}$

$m = \dfrac{y - b}{x}$

Therefore, $m = \dfrac{y - b}{x}$.

22. $L = X - \dfrac{Z}{N}$

$\dfrac{Z}{N} = X - L$

$N \cdot \dfrac{Z}{N} = N(X - L)$

$Z = N(X - L)$

Therefore, $Z = N(X - L)$

23. $V_1 = V_2 - \dfrac{n}{t}$

$\dfrac{n}{t} = V_2 - V_1$

$$\cancel{t}^{1}\frac{n}{\cancel{t}_{1}} = t(V_2 - V_1)$$

$n = t(V_2 - V_1)$

$$\frac{n}{V_2 - V_1} = \frac{t(\cancel{V_2 - V_1}^{1})}{\cancel{V_2 - V_1}_{1}}$$

Therefore, $t = \dfrac{n}{V_2 - V_1}$.

24. $M = \dfrac{2T}{g + a}$

$$(g + a)\,M = (\cancel{g + a}^{1})\,\frac{2T}{\cancel{g + a}_{1}}$$

$gM + aM = 2T$

$aM = 2T - gM$

$$\frac{\cancel{a}^{1}\cancel{M}}{\cancel{M}_{1}} = \frac{2T - gM}{M}$$

Therefore, $a = \dfrac{2T - gM}{M}$

II   25. $A = \dfrac{P - D}{F}$

$AF = \dfrac{1}{\cancel{F}} \dfrac{P - D}{\cancel{F}}_1$

$AF = P - D$

$D = P - AF$

Therefore, $D = P - AF$.

26. $i = \dfrac{d}{1 - nd}$

$(1 - nd)\, i = (\cancel{1 - nd})^1 \dfrac{d}{\cancel{1 - nd}_1}$

$i - ndi = d$

$-ndi = d - i$

$\dfrac{\cancel{-din}^1}{\cancel{-di}_1} = \dfrac{d - i}{-di}$

$n = \dfrac{(d - i)\,(-1)}{-di(-1)}$

Therefore, $n = \dfrac{i - d}{di}$.

27. $C = a\,(r_1 + 2r_2)$

$C = ar_1 + 2ar_2$

$-ar_1 = 2ar_2 - C$

$\dfrac{\cancel{-ar_1}^1}{\cancel{-a}_1} = \dfrac{2ar_2 - C}{-a}$

$r_1 = \dfrac{(2ar_2 - C)\,(-1)}{-a(-1)}$

Therefore, $r_1 = \dfrac{C - 2ar_2}{a}$

28. $A = P\,(1 + in)$

$A = P + Pin$

$-Pin = P - A$

$\dfrac{\cancel{-Pin}^1}{\cancel{-Pi}_1} = \dfrac{P - A}{-Pi}$

$n = \dfrac{(P - A)\,(-1)}{-Pi(-1)}$

Therefore, $n = \dfrac{A - P}{Pi}$.

29. $f = \dfrac{P_1 + P_2}{d(v_1 + v_2)}$

$d\,(v_1 + v_2)\,f = \cancel{d(v_1 + v_2)}^1 \dfrac{P_1 + P_2}{\cancel{d(v_1 + v_2)}_1}$

$dfv_1 + dfv_2 = P_1 + P_2$

$dfv_1 = P_1 + P_2 - dfv_2$

$\dfrac{\cancel{dfv_1}^1}{\cancel{df}_1} = \dfrac{P_1 + P_2 - dfv_2}{df}$

Therefore, $v_1 = \dfrac{P_1 + P_2 - dfv_2}{df}$

30. $d = 2d_0 + t\,(2v - at)$

$d = 2d_0 + 2tv - at^2$

$at^2 = 2d_0 + 2tv - d$

$\dfrac{\cancel{at^2}^1}{\cancel{t^2}_1} = \dfrac{2d_0 + 2tv - d}{t^2}$

Therefore, $a = \dfrac{2d_0 + 2tv - d}{t^2}$

Answers to Assignments

I   1. 6   3. 9.95   5. 5

II   7. $y = -x$ 9. $x = \frac{3}{2} - \frac{3}{4}y$ or $x = \frac{6 - 3y}{4}$ 11. $y = \frac{x}{6} + \frac{1}{4}$ or $y = \frac{2x + 3}{12}$

III   13. $V = \frac{nRT}{P}$ 15. $T_2 = \frac{P_2V_2T_1}{P_1V_1}$ 17. $G = \frac{T}{P(a - b)}$

19. $P = S - I$ 21. $r = \frac{(b - A)100}{b}$ 23. $z = \frac{\bar{x} - u}{\sigma}$

25. $T = \frac{aM + M_2}{2}$ 27. $S = \frac{2wh - p}{p}$ or $S = \frac{2wh}{p} - 1$ 29. $h = \frac{T}{2\pi r} - r$ or $h = \frac{T - 2\pi r^2}{2\pi r}$

31. $W = \frac{r}{R} + 2P$ or $w = \frac{r + 2PR}{R}$ 33. $r = \frac{a - s}{L - S}$

Unit 28

Solutions to Drill Exercises

1.   If $n = 0$, $P = -10(0^2) + 500(0) - 1000 = -1000$.

If $n = 10$, $P = -10(10^2) + 500(10) - 1000 = 3000$.

If $n = 20$, $P = -10(20^2) + 500(20) - 1000 = 5000$.

If $n = 25$, $P = -10(25^2) + 500(25) - 1000 = 5250$.

If $n = 30$, $P = -10(30^2) + 500(30) - 1000 = 5000$.

If $n = 40$, $P = -10(40^2) + 500(40) - 1000 = 3000$.

If $n = 50$, $P = -10(50^2) + 500(50) - 1000 = -1000$.

$$P = -10n^2 + 500n - 1000$$

n	0	10	20	25	30	40	50
P	-1000	3000	5000	5250	5000	3000	-1000

A daily production of 25 machines would yield a maximum profit.

2.

3. A   (2, 4)

B   (−3, 5)

C   (−4, −1)

D   (0, −3)

E   (6, −5)

F   (3, 0)

G   (0, 6)

H   (−1, 0)

4.

$$y = -\frac{1}{2}x$$

x	−4	−2	0	2	4
y	2	1	0	−1	−2

5.

$$y = x^2 - 3$$

x	−3	−2	−1	0	1	2	3
y	6	1	−2	−3	−2	1	6

6.

$$V = s^3$$

s	0	1	2	3
V	0	1	8	27

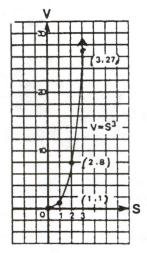

7.   (a)

$$d = \frac{1200}{V}$$

V	100	200	400	600	800	1000	1200
d	12	6	3	2	1.5	1.2	1

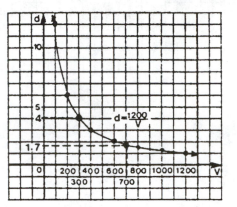

(b)   When the volume is 300 litres, the density is 4 grams/litre.

When the volume is 700 litres, the density is approximately 1.7 grams/litre.

8.   (a)

$$P = \frac{V^2}{1000}$$

V	0	20	40	60	80	100
P	0	.4	1.6	3.6	6.4	10

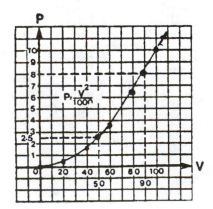

(b)   When the power is 2.5 watts, the voltage is 50 volts.
When the power is 8 watts, the voltage is 90 volts.

9. (a)

From Exercise 1. $p = -10n^2 + 500n - 1000$

n	0	10	20	25	30	40	50
p	−1000	3000	5000	5250	5000	3000	−1000

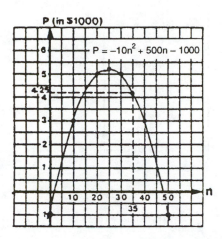

(b)   When 35 units are produced, the profit is $4250.

Answers to Assignments

1. $p = -100t^2 + 300t + 4000$

t	0	1	2	3	4	5	6	7	8
p	4000	4200	4200	4000	3600	3000	2200	1200	0

During the first day, the population increases.
The population starts to decrease after 2 days.
After 8 days, the population is extinct.

3. $y = x^2 + 1$

5.

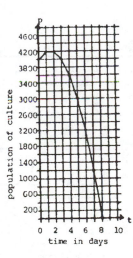

$p = -100t^2 + 300t + 4000$

After $5\frac{1}{2}$ days, the population is approximately 2600.

7.   (a)

i	0.05	0.10	0.15	0.20	0.25
P	480,000	240,000	160,000	120,000	96,000

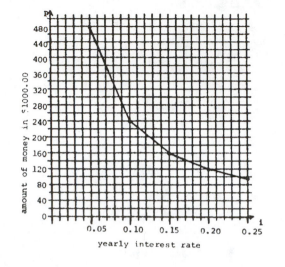

(b)

amount to invest (P)	interest rate (i)
$192,000	$12\frac{1}{2}\%$
$\simeq$ $137,000	$17\frac{1}{2}\%$
$\simeq$ $123,000	$19\frac{1}{2}\%$
$\simeq$ $106,000	$22\frac{1}{2}\%$

Unit 29

Solutions to Drill Exercises

1. $2x - 3y = 12$

x	−3	0	3
y	−6	−4	−2

2. $x + y = 6$

x	−2	0	2
y	8	6	4

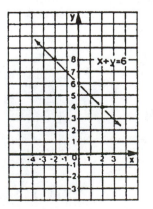

3.
$$-\frac{1}{2}p = \frac{3}{4}q - 2$$

p	−2	0	2
q	4	$2\frac{2}{3}$	$1\frac{1}{3}$

4.
$$y = 3x$$

x	−1	0	1
y	−3	0	3

5.
$$x = -2$$

x	−2	−2	−2
y	−2	0	2

6.
$$2x = 5$$

x	$\frac{5}{2}$	$\frac{5}{2}$	$\frac{5}{2}$
y	−2	0	2

7.
$$y = -3$$

x	−2	0	2
y	−3	−3	−3

8.
$$8y = 20$$

x	−3	0	3
y	$2\frac{1}{2}$	$2\frac{1}{2}$	$2\frac{1}{2}$

9. $-2y = 10x$

x	-1	0	1
y	5	0	-5

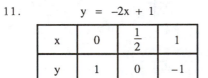

10. $2y = 3x - 6$

x	0	2	1
y	-3	0	$-\dfrac{3}{2}$

11. $y = -2x + 1$

x	0	$\dfrac{1}{2}$	1
y	1	0	-1

12. $0.75x - 1.5y = 0.3$

x	0	0.4	1
y	-0.2	0	0.3

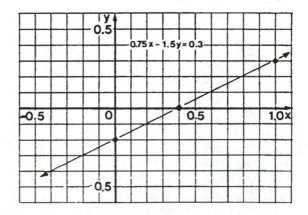

13. $\frac{5}{3}a + \frac{2}{5}j = 1$

a	0	$\frac{3}{5}$	1
j	$\frac{5}{2}$	0	$-\frac{5}{3}$

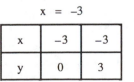

14. $3x + 2y = 6$

x	0	2
y	3	0

$m = \dfrac{y_2 - y_1}{x_2 - x_1} = \dfrac{0 - 3}{2 - 0} = -\dfrac{3}{2}$

Hence, the slope of $3x + 2y = 6$ is $-\dfrac{3}{2}$

15. $y - x = 0$

x	0	1
y	0	1

$m = \dfrac{1 - 0}{1 - 0} = 1$

Hence, the slope of $y - x = 0$ is 1.

16. $4x - 2y = 1$

x	0	$\frac{1}{4}$
y	$-\frac{1}{2}$	0

$m = \dfrac{0 - \left(-\dfrac{1}{2}\right)}{\dfrac{1}{4} - 0} = \dfrac{\dfrac{1}{2}}{\dfrac{1}{4}} = \dfrac{1}{2} \times \dfrac{4}{1} = 2$

Hence, the slope of $4x - 2y = 1$ is 2.

17. $y = 4$

x	0	2
y	4	4

Hence, the slope of $y = 4$ is 0.

(Remember $y = 4$ is the same as $0x + y = 4$; hence, for any value of x, $y = 4$.)

$m = \dfrac{4 - 4}{2 - 0} = \dfrac{0}{2} = 0$

18. $x = -3$

x	-3	-3
y	0	3

Hence, the slope of $x = -3$ is undefined.

(Remember $x = -3$ is the same as $x + 0y = -3$; hence, for any value of y, $x = -3$.)

$m = \dfrac{3 - 0}{(-3) - (-3)} = \dfrac{3}{-3 + 3} = \dfrac{3}{0}$ (undefined)

19. $y = 4 - \frac{1}{3}x$

x	0	12
y	4	0

$m = \frac{0-4}{12-0} = \frac{-4}{12} = -\frac{1}{3}$

Hence, the slope of $y = 4 - \frac{1}{3}x$ is $-\frac{1}{3}$.

20. $x + 2y = 0$

$2y = -x$

$y = -\frac{1}{2}x$

Hence, the slope of $x + 2y = 0$ is $-\frac{1}{2}$ and the coordinates of the y-intercept are $(0, 0)$.

21. $\frac{3}{5}R_1 - \frac{1}{4}R_2 = 1$

$-\frac{1}{4}R_2 = -\frac{3}{5}R_1 + 1$

$(-4)\left(-\frac{1}{4}R_2\right) = (-4)\left(-\frac{3}{5}R_1 + 1\right)$

$R_2 = \frac{12}{5}R_1 - 4$

Hence, the slope of $\frac{3}{5}R_1 - \frac{1}{4}R_2 = 1$ is $\frac{12}{5}$ and the coordinates of the R_2-intercept are $(0, -4)$.

22. $0.06a + 0.02j = 0.12$

$0.02j = -0.06a + 0.12$

$\frac{0.02j}{0.02} = \frac{-0.06a + 0.12}{0.02}$

$j = \frac{-0.06a}{0.02} + \frac{0.12}{0.02}$

$= -3a + 6$

Hence, the slope of $0.06a + 0.02j = 0.12$ is -3 and the coordinates of the j-intercept are $(0, 6)$.

23. $2I - 4E = 0$

$-4E = -2I$

$\frac{-4E}{-4} = \frac{-2I}{-4}$

$E = \frac{1}{2}I$

Hence, the slope of $2I - 4E = 0$ is $\frac{1}{2}$ and the coordinates of the E-intercept are $(0, 0)$.

24.   The coordinates of two points on the line are $(0, 0)$ and $(-2, 1)$.

$\therefore m = \frac{y_2 - y_1}{x_2 - x_1} = \frac{1 - 0}{-2 - 0} = -\frac{1}{2}$

The y-intercept is 0.

Hence, substituting $m = -\frac{1}{2}$ and $b = 0$ into $y = mx + b$, we have $y = -\frac{1}{2}x$.

25.   The coordinates of the x- and y-intercepts are (–2, 0) and (0, 10).

$$\therefore m = \frac{10 - 0}{0 - (-2)} = \frac{10}{2} = 5$$

The y-intercept is 10.

Hence, the equation is $y = 5x + 10$.

26.   E = 3I

I	0	2
E	0	6

(a)   The E-intercept is 0 which means that when the current I is zero, the voltage E is zero.

(b)   The slope is 3. It represents the resistance in ohms and indicates that for an increase of 1 ampere in the current, the voltage has to increase by 3 volts.

27.   N = 0.7x + 4800   (*Note*: x cannot be negative)

x	0	1000
N	4800	5500

(a)   The N-intercept is 4800 which means that 4800 is the number of orders received without distributing any catalogues.

(b)   The slope is 0.7. It means that for every catalogue sent out, the company expects to receive 0.7 orders, or for every 1000 catalogues sent out, the company expects 700 orders.

Answers to Assignments

1.

$$3x + 4y = -5$$

3.

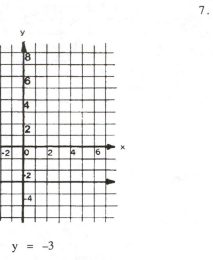

$$y = -8x$$

5.

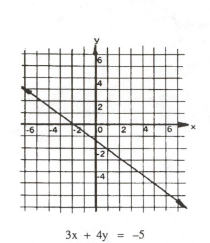

$$y = -3$$

7.

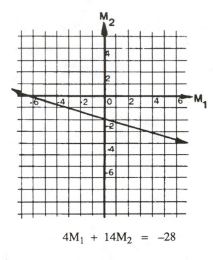

$$4M_1 + 14M_2 = -28$$

9.

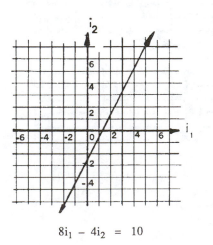

$$8i_1 - 4i_2 = 10$$

11.

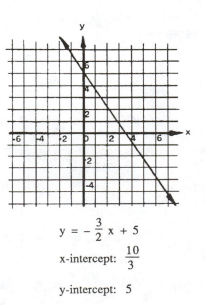

$$y = -\frac{3}{2}x + 5$$

x-intercept: $\frac{10}{3}$

y-intercept: 5

13.

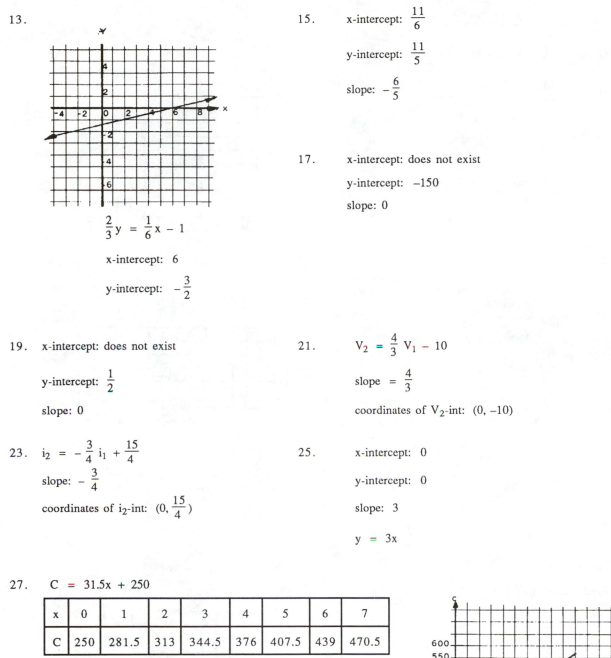

$$\frac{2}{3}y = \frac{1}{6}x - 1$$

x-intercept: 6

y-intercept: $-\frac{3}{2}$

15. x-intercept: $\frac{11}{6}$

y-intercept: $\frac{11}{5}$

slope: $-\frac{6}{5}$

17. x-intercept: does not exist

y-intercept: -150

slope: 0

19. x-intercept: does not exist

y-intercept: $\frac{1}{2}$

slope: 0

21. $V_2 = \frac{4}{3}V_1 - 10$

slope $= \frac{4}{3}$

coordinates of V_2-int: $(0, -10)$

23. $i_2 = -\frac{3}{4}i_1 + \frac{15}{4}$

slope: $-\frac{3}{4}$

coordinates of i_2-int: $(0, \frac{15}{4})$

25. x-intercept: 0

y-intercept: 0

slope: 3

$y = 3x$

27. $C = 31.5x + 250$

x	0	1	2	3	4	5	6	7
C	250	281.5	313	344.5	376	407.5	439	470.5

x	8	9	10
C	502	533.5	565

(a) C-intercept: 250

(b) slope: 31.5

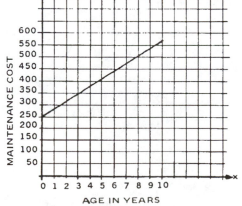

Unit 30

Solutions to Drill Exercises

1. The coordinates of the point of
    intersection are (4, 0).

    Hence, x = 4 and y = 0 is the
    solution of the system.

    Check:

    $$x - 2y = 4 \quad \text{and} \quad x + 4y = 4$$
    $$4 - 2(0) = 4 \qquad\quad 4 + 4(0) = 4$$

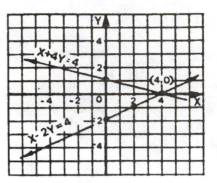

2. The coordinates of the point of
    intersection are (3, –2).

    Hence, x = 3 and y = –2 is the
    solution of the system.

    Check:

    $$x - y = 5 \quad \text{and} \quad 2x + y = 4$$
    $$3 - (-2) = 5 \qquad\quad 2(3) + (-2) = 4$$
    $$3 + 2 = 5 \qquad\qquad\quad 6 - 2 = 4$$

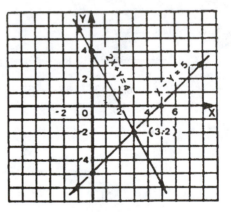

3. The coordinates of the point of
    intersection are (–1, 2).

    Hence, x = –1 and y = 2 is the
    solution of the system.

    Check:

    $$y = x + 3 \quad \text{and} \quad x + 3y = 5$$
    $$2 = -1 + 3 \qquad\quad -1 + 3(2) = 5$$
    $$-1 + 6 = 5$$

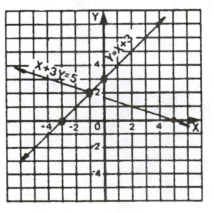

4. $n = 2m$

$7m = 35 + n$

Substitute $2m$ for n into $7m = 35 + n$.

∴   (a) $7m = 35 + (2m)$ (b)   Solve for n

$5m = 35$ $n = 2m$

$m = 7$ $n = 2(7)$

$= 14$

Check: $n = 2m$ and $7m = 35 + n$

$14 = 2(7)$ $7(7) = 35 + 14$

$49 = 49$

Hence, the solution is $m = 7$ and $n = 14$.

5. $R_1 = 21 - 6R_2$

$\dfrac{R_1}{3} = R_2 + 1$

Substitute $21 - 6R_2$ for R_1 into $\dfrac{R_1}{3} = R_2 + 1$.

∴   (a) $\dfrac{21 - 6R_2}{3} = R_2 + 1$ and   (b)   Solve for R_1.

$21 - 6R_2 = 3R_2 + 3$ $R_1 = 21 - 6R_2$

$-9R_2 = -18$ $= 21 - 6(2)$

$R_2 = 2$ $= 9$

Check:

$R_1 = 21 - 6R_2$ and $\dfrac{R_1}{3} = R_2 + 1$

$9 = 21 - 6(2)$ $\dfrac{9}{3} = 2 + 1$

Hence, the solution is $R_1 = 9$ and $R_2 = 2$.

6. $0.1x + 0.3y = 0.9$ $10(0.1x + 0.3y) = 10(0.9)$ or $x + 3y = 9$

$0.5x + 0.7y = 3.3$ $10(0.5x + 0.7y) = 10(3.3)$ $5x + 7y = 33$

Since $x + 3y = 9$, we substitute $-3y + 9$ for x into $5x + 7y = 33$.

(a) $5(-3y + 9) + 7y = 33$ (b)   Solve for x.

$-15y + 45 + 7y = 33$ $x + 3y = 9$

$-8y = -12$ $x + 3\left(\dfrac{3}{2}\right) = 9$

$y = \dfrac{3}{2}$ $x + \dfrac{9}{2} = 9$

$x = 9 - \dfrac{9}{2}$

$x = \dfrac{9}{2}$

6.   *(continued)*

Check:

$$0.1x + 0.3y = 0.9 \qquad \text{and} \qquad 0.5x + 0.7y = 3.3$$

$$\frac{1}{10}\left(\frac{9}{2}\right) + \frac{3}{10}\left(\frac{3}{2}\right) = \frac{9}{10} \qquad\qquad\qquad \frac{5}{10}\left(\frac{9}{2}\right) + \frac{7}{10}\left(\frac{3}{2}\right) = \frac{33}{10}$$

$$\frac{9}{20} + \frac{9}{20} = \frac{9}{10} \qquad\qquad\qquad\qquad \frac{45}{20} + \frac{21}{20} = \frac{33}{10}$$

$$\frac{18}{20} = \frac{9}{10} \qquad\qquad\qquad\qquad\qquad \frac{66}{20} = \frac{33}{10}$$

Hence, the solution is $x = \frac{9}{2}$ and $y = \frac{3}{2}$.

7. $V_1 = 2(V_2 - 5) \qquad \rightarrow \qquad V_1 = 2V_2 - 10$

$\underline{4V_1 - 8V_2 = -40}$

Substitute $2V_2 - 10$ for V_1 into $4V_1 - 8V_2 = -40$

(a) $4(2V_2 - 10) - 8V_2 = -40$

$8V_2 - 40 - 8V_2 = -40$

$-40 = -40$ (This is true.)

This is a dependent system which has an infinite number of solutions.
No unique solution can be found.

8.   (a) $3x - 8y = -1$

$\underline{9x - 8y = 1} \qquad \text{(Subtract)}$

$-6x + 0 = -2$

$x = \frac{1}{3}$

(b)   Solve for y. $3x - 8y = -1$

$3\left(\frac{1}{3}\right) - 8y = -1$

$1 - 8y = -1$

$-8y = -2$

$y = \frac{1}{4}$

Check:

$$3x - 8y = -1 \qquad \text{and} \qquad 9x - 8y = 1$$

$$3\left(\frac{1}{3}\right) - 8\left(\frac{1}{4}\right) = -1 \qquad\qquad 9\left(\frac{1}{3}\right) - 8\left(\frac{1}{4}\right) = 1$$

$$1 - 2 = -1 \qquad\qquad\qquad\qquad 3 - 2 = 1$$

Hence, the solution is $x = \frac{1}{3}$ and $y = \frac{1}{4}$.

9. (a) $\begin{array}{ll} 3a - 6j &= -63 \\ 9a - 5j &= -85 \end{array}$ $\rightarrow$ $\begin{array}{ll} 3(3a - 6j) &= 3(-63) \\ 9a - 5j &= -85 \end{array}$ or $\begin{array}{ll} 9a - 18j &= -189 \\ 9a - 5j &= -85 \\ \hline 0 - 13j &= -104 \\ j &= 8 \end{array}$

(b) Solve for a. $\begin{array}{ll} 3a - 6j &= -63 \\ 3a - 6(8) &= -63 \\ 3a - 48 &= -63 \\ 3a &= -15 \\ a &= -5 \end{array}$

Check:

$\begin{array}{ll} 3a - 6j &= -63 \\ 3(-5) - 6(8) &= -63 \\ -15 - 48 &= -63 \end{array}$ and $\begin{array}{ll} 9a - 5j &= -85 \\ 9(-5) - 5(8) &= -85 \\ -45 - 40 &= -85 \end{array}$

10. (a) $\begin{array}{ll} 5m + 8n &= 19 \\ -2m + 5n &= 17 \end{array}$ $\rightarrow$ $\begin{array}{ll} 2(5m + 8n) &= 2(19) \\ 5(-2m) + 5n) &= 5(17) \end{array}$ or $\begin{array}{ll} 10m + 16n &= 38 \\ -10m + 25n &= 85 \\ \hline 0 + 41n &= 123 \\ n &= 3 \end{array}$ (Add)

(b) Solve for m. $\begin{array}{ll} 5m + 8n &= 19 \\ 5m + 8(3) &= 19 \\ 5m + 24 &= 19 \\ 5m &= -5 \\ m &= -1 \end{array}$

Check:

$\begin{array}{ll} 5m + 8n &= 19 \\ 5(-1) + 8(3) &= 19 \\ -5 + 24 &= 19 \end{array}$ and $\begin{array}{ll} -2m + 5n &= 17 \\ -2(-1) + 5(3) &= 17 \\ 2 + 15 &= 17 \end{array}$

11. $\begin{array}{ll} \dfrac{3}{5}R_1 - \dfrac{6}{5}R_2 &= 3 \\ 7R_1 + 11R_2 &= 35 \end{array}$ $5\left(\dfrac{3}{5}R_1 - \dfrac{6}{5}R_2\right) = 5(3)$ $7R_1 + 11R_2 = 35$ or $\begin{array}{ll} 3R_1 - 6R_2 &= 15 \\ 7R_1 + 11R_2 &= 35 \end{array}$

(a) $\begin{array}{ll} 11(3R_1 - 6R_2) &= 11(15) \\ 6(7R_1 + 11R_2) &= 6(35) \end{array}$ or $\begin{array}{ll} 33R_1 - 66R_2 &= 165 \\ 42R_1 + 66R_2 &= 210 \\ \hline 75R_1 + 0 &= 375 \\ R_1 &= 5 \end{array}$

11.   *(continued)*

(b)   Solve for R_2.

$$\frac{3}{5}R_1 - \frac{6}{5}R_2 = 3$$

$$\frac{3}{5}(5) - \frac{6}{5}R_2 = 3$$

$$3 - \frac{6}{5}R_2 = 3$$

$$-\frac{6}{5}R_2 = 0$$

$$R_2 = 0$$

Check:

$$\frac{3}{5}R_1 - \frac{6}{5}R_2 = 3 \qquad 7R_1 + 11R_2 = 35$$

$$\frac{3}{5}(5) - \frac{6}{5}(0) = 3 \qquad 7(5) + 11(0) = 35$$

$$3 - 0 = 3 \qquad 35 + 0 = 35$$

Hence, the solution is $R_1 = 5$ and $R_2 = 0$.

12.   The solution is $x = -10$ and $y = 6$. The system is consistent.

13.   This is a dependent system which has an infinite number of solutions, but no unique one.

14.   The solution is $V_1 = 2$ and $V_2 = -1$. The system is consistent.

15.   This is an inconsistent system which has no solution.

16.   The solution is $R_1 = \frac{3}{2}$ and $R_2 = -\frac{1}{3}$. This system is consistent.

17.   The solution is $M = 7$ and $N = 30$. This system is consistent.

18.   Let $n =$ the larger number, and19.

 $m =$ the smaller number.

Hence, $n - m = 12$

     $n + m = 28$

The solution of this system is $n = 20$ and $m = 8$.

Therefore, the larger number is 20 and the smaller is 8.

19.   Let $n =$ the larger number, and

 $m =$ the smaller number.

Hence, $n + m = 76$

     $n = 3m$

The solution of this system is $n = 57$ and $n = 19$.

Therefore, the larger number is 57 and the smaller is 19.

20.   Let $x =$ cost of one of the machines, and

 $y =$ cost of the other machine.

Hence, $x + y = 80\ 000$

     $x = 2y - 10\ 000$

The solution of this system is $x = 50\ 000$ and $y = 30\ 000$.

The cost of one machine is $50 000 and the cost of the other is $30 000.

21. Let x = the number of litres of 30% solution, and

   y = the number of litres of 80% solution.

Hence, $x + y = 10$

                    $0.3x + 0.8y = 0.6(10)$

The solution of this system is $x = 4$ and $y = 6$.

Hence, 4 litres of the 30% solution and 6 litres of the 80% solution are required to produce 10 litres of a 60% hydrochloric acid solution.

22. (a) $15 = 2.0I_1 + 1.5(I_1 - I_2)$ $15 = 2.0I_1 + 1.5I_1 - 1.5I_2$
                                           or
            $0 = 3I_2 + 1.5(I_2 - I_1)$ $0 = 3I_2 + 1.5I_2 - 1.5I_1$

            $\therefore\ 3.5I_1 - 1.5I_2 = 15$

            $-1.5I_1 + 4.5I_2 = 0$ or $-1.5I_1 = -4.5I_2$

                                                   $I_1 = 3I_2$

      Substitute $3I_2$ for I_1 into $3.5I_1 - 1.5I_2 = 15$.

            $\therefore\ 3.5(3I_2) - 1.5I_2 = 15$

            $10.5I_2 - 1.5I_2 = 15$

            $9I_2 = 15$

            $I_2 = \dfrac{5}{3}$

   (b)   Solve for I_1 $I_1 = 3I_2$

                          $= 3\left(\dfrac{5}{3}\right)$

                          $= 5$

      Check: $15 = 2.0I + 1.5(I_1 - I_2)$ and $0 = 3I_2 + 1.5(I_2 - I_1)$

                $15 = 2.0(5) + 1.5\left(5 - \dfrac{5}{3}\right)$ and $0 = 3\left(\dfrac{5}{3}\right) + 1.5\left(\dfrac{5}{3} - 5\right)$

                $15 = 10 + 5$ $0 = 5 - 5$

Answers to Assignments

1. $x = 1, y = -2$ consistent   3.   no solution   inconsistent

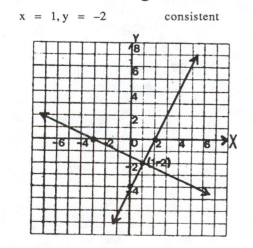

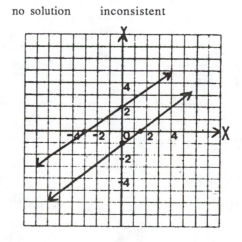

5. $d = -1$, $t = 5$ 7. $x = \dfrac{1}{2}$, $y = \dfrac{2}{3}$ 9. $V_1 = 3$, $V_2 = \dfrac{3}{2}$

11. $m = 600$, $n = 30$ 13. $a = 7$, $j = 11$ 15. 37 & 43

17. 30 – $20 cheques; 18 – $50 cheques   19. $105 000 at 18%, $75 000 at 12%

21. 36 L of 20% solution, 24 L of 45% solution   23. Model A component costs $325
                                                          Model B component costs $375

Unit 31

Solutions to Drill Exercises

I.

	Exact Numbers			Approximate Numbers
	Pure	Counting	Defined	Measurement
1.	118.25, 0.25, 29.562 5			
2.			1, $\dfrac{1}{60}$	
3.		21		11:00 h
4.		50		36 cm
5.				120 km, 16 ℓ
6			1000, 1	
7.	156.25, 12.5			
8.				67° 50′, 162 V
9.				2.5 t, 75°
10.				16:00 h, –18°C
11.		20		650 ℓ, 1 h
12.	$\dfrac{3}{4}$, $\dfrac{1}{6}$	252		

II   (a)

	Number	Significant Digits
13.	7000	one
14.	0.000 24	two
15.	52 001	five
16.	95.00	four
17.	800 00$\tilde{0}$	six
18.	0.100	three
19.	29 000	two
20.	446 020 0	seven

(b)   0.000 24 is the most precise number and
7000 or 29 000 are the least precise numbers

III

	Calculated Result	Least Precise Number	Rounded Answer
21.	17.319 182	0.001	17.319
22.	3 642.281	125	3642
23.	38 850	200	38 800
24.	34.082 4	123.01	34.08
25.	686 400	250 000	690 000
26.	0.046 01	0.003 8	0.046 0

IV

	Step 1 Result of calculation	*Step 2* Number with fewest significant digits	*Step 3* Number of significant digits in Step 2 number	*Step 4* Rounded answer of calculation
27.	94.50	100	one	90
28.	290.015	10.000	five	290.02
29.	1162	8.3 and 140	two	1200
30.	10	100.00	five	10.000
31.	0.734 846 9	0.54	two	0.73
32.	1.6	2.56	three	1.60
33.	13.5	0.30	two	14
34.	0.001 062 555 5	90	one	0.001
35.	1 653.480 3	112.49	five	1 653.5
36.	285.61	16.9	three	286
37.	447 561	669	three	448 000
38.	0.09	0.300 0	four	0.090 00

Answers to Assignment

I   1. A, A   3. E, E   5. E, A   7. E, E, E   9. E, E

II 11. 3   13. 2   15. 7   17. 4   19. 3

Most precise #19
Least precise #14

III 21. 461.4   461   460   500

23. 1.342   1.34   1.3   1

25. 81 750   81 800   82 000   80 000

IV 27. 14.00   29. 0.3002   31. 59.65

V 33. 1000   35. 100.000 0   37. 454

39. 0.003 60   41. 8000

Unit 32

Solutions to Drill Exercises

I   1. $\angle B = 31°$ 2. $\angle P = 88°$

II   3. acute   4. straight   5. right   6. obtuse

III 7. (a) scalene, none of its sides or angles are equal   8. (a) right, $\angle S = 90°$

     (b) $\angle B = 92°$ (b) $\angle Q = 61°$

     9. (a) isosceles, two sides and two angles are equal   10. (a) equilateral, three sides and three angles are equal

     (b) $\angle T = 55°, \angle M = 55°$ (b) $\angle F = 60°, \angle E = 60°, \angle G = 60°$

IV   11. (a) Using the Pythagorean Theorem, we have   (b) $\angle A + \angle B = 90°$ (complementary angles)

$$c^2 = a^2 + b^2$$
$$= (14)^2 + (8)^2$$
$$= 196 + 64$$
$$= 260$$
$$c = \sqrt{260}$$
$$= 16.124\ 5$$
$$= 16 \text{ mm}$$

(rounded to the nearest whole number)

The length of side $c = 16$ mm.

$$\angle B = 90° - \angle A$$
$$= 90° - 60°$$
$$= 30°$$

The size of $\angle B = 30°$.

12. (a) Using the Pythagorean Theorem, we have   (b) $\angle D + \angle E = 90°$

$$f^2 = e^2 + d^2$$

Since we are interested in the length of side d, we rearrange the formula by subtracting e^2 from both sides.

$$f^2 - e^2 = e^2 + d^2 - e^2$$
$$f^2 - e^2 = d^2$$
$$\text{or} \quad d^2 = f^2 - e^2$$
$$= (10.9)^2 - (6.3)^2$$
$$= 118.81 - 39.69$$
$$= 79.12$$
$$d = \sqrt{79.12}$$
$$= 8.894\ 9$$
$$= 8.9 \text{ cm}$$

(rounded to 2 significant digits)

The length of side $d = 8.9$ cm.

$$\angle D = 90° - \angle E$$
$$= 90° - 35°$$
$$= 90° - 35°$$
$$= 90° - 35°$$

The size of $\angle D = 55°$.

Answers to Assignment

I   1.   complementary

3.   complementary

II   5.   acute

7.   right

III   9.   isosceles and
right

Property: Two equal sides, two equal angles, and the
third angle equals 90°.

11.   equilateral

Property: Three equal sides and three equal angles.

IV   13. $\angle A = \angle C = 45°$

15. $\angle E = \angle F = \angle G = 60°$

V   17.   r = 6.7 m

$\angle S = 40°$

19.   f = 3.89 km

$\angle E = \angle G = 45°$

Unit 33

Solutions to Drill Exercises

I.   1.   C   = $2\pi r$

= 2π (9.50)

=   59.690 26

=   59.7 cm (rounded)

A   = πr^2

= $\pi (9.50)^2$

=   283.528 74

=   284 cm^2 (rounded)

2.   P   = $2\ell + 2w$

=   2(31.45) + 2(15.26)

=   62.9 + 30.52

=   93.42 m

A   = ℓw

=   (31.45) (15.26)

=   479.927

=   479.9 m^2 (rounded)

3.   P   =   25.0 + 15.0 + 12.0 + 12.6

=   64.6 m

A   = $\left(\dfrac{b_1 + b_2}{2}\right) h$

= $\left(\dfrac{25.0 + 12.0}{2}\right)$ (12.0)

=   (18.5) (12.0)

=   222 m^2

4.   P   =   2s + 2b

=   2(175) + 2(316)

=   350 + 632

=   982 mm

A   =   bh

=   (316) (90.5)

=   28 598

=   28 600 mm^2 (rounded)

I   5. $P = 4s$ $A = s^2$

         $= 4(0.960)$ $= (0.960)^2$

         $= 3.84 \text{ km}$ $= 0.9216$

                                         $= 0.922 \text{ km}^2 \text{ (rounded)}$

    6. $P = 12.5 + 4.60 + 14.7$ $A = \frac{1}{2}bh$

         $= 31.8 \text{ cm}$ $= \frac{1}{2}(14.7)(3.70)$

                                         $= 27.195$

                                         $= 27.2 \text{ cm}^2 \text{ (rounded)}$

    7. $P = 2(6.20) + \pi(3.46)$ $A = \text{area rectangle} + \text{area circle}$

         $= 23.269\ 911$ $= \ell w + \pi r^2$

         $= 23.3 \text{ m (rounded)}$ $= (6.20)(3.46) + \pi(1.73)^2$

                                         $= 30.854\ 473$

                                         $= 30.9 \text{ m}^2 \text{ (rounded)}$

II   8. $V = \ell wh$ 9. $V = \pi r^2 h$

          $= (9.2)(7.8)(.065)$ $= \pi(2.1)^2(3.5)$

          $= 4.664\ 4$ $= 48.490\ 483$

          $= 4.7 \text{ m}^3$ $= 48 \text{ m}^3$

   ∴   The roof holds 4.7 cubic metres of water.   ∴   The volume of the tank is 48 cubic metres.

    10. (a) $V = \text{area of triangular base} \times \text{height of prism}$

               $= \frac{1}{2}(10.0)(17.3) \times 14.5$

               $= 1254.25$

               $= 1250 \text{ cm}^3 \text{ (rounded)}$

         (b) $\text{lateral area} = 14.5(17.3 + 10.0 + 20.0)$

               $= 14.5(47.3)$

               $= 685.85$

               $= 686 \text{ cm}^2 \text{ (rounded)}$

         (c) $\text{total surface area} = 2(\text{area of base}) + \text{lateral area}$

               $= 2\left(\frac{1}{2}\right)(10.0)(17.3) + 686$

               $= 173 + 686$

               $= 859 \text{ cm}^2$

II   11.   (a) $V = \frac{4}{3}\pi r^3$

$= \frac{4}{3}\pi (6.30)^3$

$= 1047.394\ 4$

$= 1050\ m^3$ (rounded)

(b) $SA = 4\pi r^2$

$= 4\pi (6.30)^2$

$= 498.759\ 25$

$= 499\ m^2$ (rounded)

12. $SA = 2\pi r^2 + 2\pi rh$

$= 2\pi (8.00)^2 + 2\pi (8.00)(12.00)$

$= 402.123\ 86 + 603.185\ 79$

$= 1005.309\ 6$

$= 1010\ m^2$ (rounded)

13. $V = \frac{1}{3}\pi r^2 h$

$= \frac{1}{3}\pi (2.5)^2 (7.2)$

$= 47.123\ 89$

$= 47\ cm^3$ (rounded)

$SA = \pi rs + \pi r^2$

$= \pi (2.5)(7.6) + \pi (2.5)^2$

$= 59.690\ 26 + 19.634\ 954$

$= 79.325\ 214$

$= 79\ cm^2$ (rounded)

$S = \sqrt{(2.5)^2 + (7.2)^2}$

$= 7.621\ 679\ 6$

$= 7.6\ cm$

14. V = area of trapezoidal end x length

$= \left(\frac{3.00 + 6.00}{2}\right)(3.20)\ x\ 11.5$

$= 165.6$

$= 166\ cm^3$ (rounded)

$SA = 2\left(\frac{3.00 + 6.00}{2}\right)(3.20) + (11.5)(6.00 + 3.50 + 3.00 + 3.50)$

$= 28.8 + 184$

$= 212.8$

$= 213\ cm^2$ (rounded)

Answers to Assignment

I   1.   perimeter = 3.36 km

area = 0.692 km² or 692 000 m²

3.   perimeter = 131.0 cm

area = 1073 cm²

I   5.   perimeter = 19 cm

area = 11 cm^2

7.   perimeter = 35 m

area = 45 m^2

II   9.   Width = 5.0 cm

11.   Volume of metal used = 6980 cm^3.

13.   Volume = 4564 m^3 and will hold 4 564 000 L of water.

15.   Volume = 210 m^3

Total Surface Area = 220 m^2

Unit 34

Solutions to Drill Exercises

I   1.   Hypotenuse

2.   Adjacent side

3.   Adjacent side

II   4. $\dfrac{12}{37}$

5. $\dfrac{35}{37}$

6. $\dfrac{12}{35}$

7. $\dfrac{35}{37}$

8. $\dfrac{35}{12}$

9. $\dfrac{12}{37}$

III   10.   *Step 1*

Step 2   Known parts: $\angle A$ = 32°

a = 21 m (opposite side)

Unknown part: c = ? (hypotenuse)

We use $\sin A$ = $\dfrac{a}{c}$.

Step 3 $\sin 32° = \dfrac{21}{c}$

$0.5299 = \dfrac{21}{c}$

$0.5299 \times c = \dfrac{21}{c} \times c$

$\dfrac{0.5299c}{0.5299} = \dfrac{21}{0.5299}$

$\therefore c = 39.63$

Therefore, the length of side c = $4\tilde{0}$ m (rounded to two significant digits)

III   11.   *Step 1*

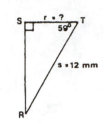

Step 2   Known part: $\angle T$ =   59°

s   =   12 mm (hypotenuse)

Unknown part: r =   ? (side adjacent to $\angle T$)

We use $cos\ T$ = $\dfrac{r}{s}$.

Step 3 $\cos 59°$ = $\dfrac{r}{12}$

0.5150   = $\dfrac{r}{12}$

0.5150 x 12   = $\dfrac{r}{12}$ x 12

$\therefore r$ =   6.18

Therefore, the length of side r = 6.2 mm.

12.   *Step 1*

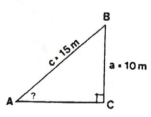

Step 2   Unknown part: $\angle A$ =   ?

Known part: a =   10 m (side opposite $\angle A$)

c =   15 m (hypotenuse)

We use $\sin A$ = $\dfrac{a}{c}$.

Step 3 $\sin A$ = $\dfrac{10}{15}$

$\sin A$ =   0.6667

$\therefore \angle A$ = $\sin^{-1}$ (0.6667)

=   41.81°

Therefore, $\angle A$ = 42° (rounded to the nearest degree).

III 13. *Step 1*

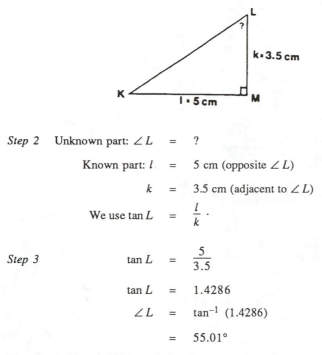

Step 2   Unknown part: $\angle L$ = ?

Known part: l = 5 cm (opposite $\angle L$)

k = 3.5 cm (adjacent to $\angle L$)

We use $\tan L = \dfrac{l}{k}$.

Step 3

$$\tan L = \frac{5}{3.5}$$

$$\tan L = 1.4286$$

$$\angle L = \tan^{-1}(1.4286)$$

$$= 55.01°$$

Therefore, $\angle L = 55°$ (rounded to the nearest degree).

Answers to Assignment

I 1. (a) $\sin Q = \dfrac{4}{5}$, $\cos Q = \dfrac{3}{5}$, $\tan Q = \dfrac{4}{3}$

     (b) $\sin S = \dfrac{3}{5}$, $\cos S = \dfrac{4}{5}$, $\tan S = \dfrac{3}{4}$

II 3.   (4.695 round to two significant digits) ∴ length of side f = 4.7 km

   5.   (58.3970 round to two significant digits) ∴ length of side c = 58 km

   7.   (37.40°) ∴ $\angle C = 37°$

Unit 35

Solutions to Drill Exercises

I   1. $\tan Q$ 2. $\csc A$ 3. $\cos M$

   4. $\sec \alpha$ 5. $\cot C$ 6. $\sin C$

II   7. $\dfrac{n}{m}$ 8. $\dfrac{4}{5}$ 9. $\dfrac{m}{r}$

   10. $\dfrac{5}{3}$ 11. $\dfrac{m}{r}$ 12. $\dfrac{4}{5}$

   13. $\dfrac{3}{4}$ 14. $\dfrac{5}{3}$ 15. $\dfrac{n}{r}$

III   16.

Angles	csc	sec	cot
0	undefined	1.0000	undefined
4°	14.3356	1.0024	14.3007
16°	3.6280	1.0403	3.4874
28°	2.1301	1.1326	1.8807
37°	1.6616	1.2521	1.3270
49°	1.3250	1.5243	0.8693
54°	1.2361	1.7013	0.7265
61°	1.1434	2.0627	0.5543
71°	1.0576	3.0716	0.3443
76°	1.0306	4.0211	0.2493
90°	1.0000	undefined	0.0000

IV   17.   (a)   36°16′

63°49′

99°65′ = 99° + 1°5′ = 100°5′

(c)   84°34′

24°26′

108°60′ = 108° + 1° = 109°

(b)   3°59′

18°57′

21°116′ = 21° + 1°56′ = 22°56′

18. $\angle L$ = $180° - (\angle L + \angle M)$

= $180° - (42°53′ + 93°41′)$

= $180° - 135°94′$

= $180° - 136°34′$

= $179°60′ - 136°34′$

= $43°26′$

V   19.   (a) $39′ = 39 \times \left(\frac{1}{60}\right)° = 0.65°$

∴ 123°39′ = 123.65°

(b) $2′ = 2 \times \left(\frac{1}{60}\right)° = 0.03°$

∴ 11°2′ = 11.03°

20.   (a) $0.86° = 0.86 \times 60′ = 51.6′$

$71.86° = 71°52′$
rounded to the nearest minute.

(b) $0.37° = 0.37 \times 60′ = 22.2′$

or $0.37° = 22′$
rounded to the nearest minute.

V 21. (a) $\angle C$ = $180° - (\angle A + \angle B)$

                = $180° - (24.8° + 123.3°)$

                = $180° - 148.1°$

                = $31.9°$

    (b) $0.9°$ = $0.9 \times 60' = 54'$

          $\therefore \angle C$ = $31°54'$

Answers to Assignment

I   1. csc Q   3. sin M   5. sec A

II   7. $\dfrac{n}{m}$ 9. $\dfrac{m}{r}$ 11. $\dfrac{3}{4}$

    13. $\dfrac{m}{n}$ 15. $\dfrac{4}{3}$

IV 17. (a) $\angle H$ = $50°45'$ V   19. (a) $\angle B$ = $47.7°$

        (b) $\angle H$ = $50.75°$ (b) $\angle B$ = $47°42'$

             $\angle F$ = $39.25°$ $\angle A$ = $73°24'$

                                                          $\angle C$ = $58°54'$

Unit 36

Solutions to Drill Exercises

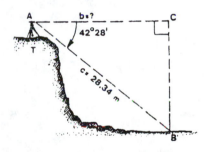

1. Step 1.   Let A be the point on top of the transit directly
              over point T. The distance AC equals the horizontal
              distance between points T and B. Draw a diagram
              and label the parts.

    Step 2.   Unknown: b   =   ? (side adjacent to $\angle$ A)

             Known: c   =   28.34 m (hypotenuse)

                  $\angle$ A   = $42°28'$

             We use cos A   = $\dfrac{b}{c}$

    Step 3. $\cos 42°28'$ = $\dfrac{b}{28.34}$

               $\cos 42.47°$ = $\dfrac{b}{28.34}$

          0.7376×28.34 = $\dfrac{b}{28.34} \times 28.34$

                  $\therefore$ b   =   20.90 m

    Step 4.   We find that the horizontal distance between point T and point B is 20.90 m.

2.   Step 1.

Step 2.   Unknown: b = ? (hypotenuse)

Known: $\angle$ F = 75°

f = 28 m (side opposite $\angle$ F)

We use sin F = $\dfrac{f}{b}$.

Step 3.   sin 75° = $\dfrac{28}{b}$

0.9659 = $\dfrac{28}{b}$

0.9659 x b = $\dfrac{28}{b}$ x b

$\dfrac{0.9659}{0.9659}$ x b = $\dfrac{28}{0.9659}$

b = 28.99 m

Step 4.   Hence, the length of the ladder for the new fire truck must be at least 29 m to reach the top of the tallest building.

3.   Step 1.

Step 2.   Unknown: $\angle$ H = ?

Known: h = 77.10 m (side opposite $\angle$ H)

r = 1000.00 m (hypotenuse)

We use sin H = $\dfrac{h}{r}$.

Step 3.   sin H = $\dfrac{77.1}{1000}$.

= 0.0771

$\therefore$ $\angle$ H = arc sin (0.0771) (or sin^{-1}, inv sin)

= 4.4219°

= 4°25′

Step 4.   Hence, the highway elevates at an angle of 4°25′.

4.   Step 1.

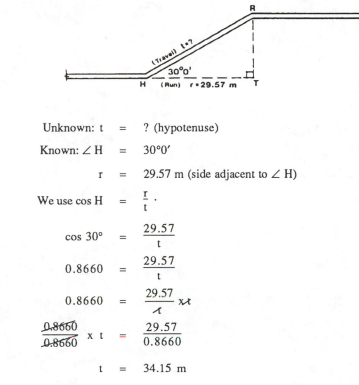

Step 2.   Unknown: t   =   ? (hypotenuse)

Known: ∠ H   =   30°0′

r   =   29.57 m (side adjacent to ∠ H)

We use cos H   = $\dfrac{r}{t}$.

Step 3.   cos 30°   = $\dfrac{29.57}{t}$

0.8660   = $\dfrac{29.57}{t}$

0.8660   = $\dfrac{29.57}{\cancel{t}}$ × t

$\dfrac{\cancel{0.8660}}{\cancel{0.8660}}$ × t   = $\dfrac{29.57}{0.8660}$

t   =   34.15 m

Step 4.   Hence, the required minimum length of pipe for the connection is 34.15 m.

5.   *Note:*   Here we will use the *given* measurement (r = 29.57 m) rather than the part
(t = 34.15 m) calculated in Exercise 4 to find the unknown part h.

Step 1.

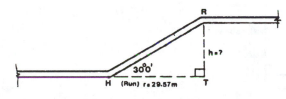

Step 2.   Unknown: h   =   ? (side opposite ∠ H)

Known: ∠ H   =   30°0′

r   =   29.57 m (side adjacent to ∠ H)

We use tan H   = $\dfrac{h}{r}$.

Step 3.   tan 30°0′   = $\dfrac{h}{29.57}$

0.5774   = $\dfrac{h}{29.57}$

0.5774 × 29.57   = $\dfrac{h}{\cancel{29.57}}$ × $\cancel{29.57}$

h   =   17.07 m

Step 4.   Hence, the vertical elevation between the two pipelines is 17.07 m.

6.   Step 1.

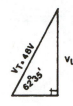

Step 2.   Unknown: V_L =   ? (side opposite $\angle \phi$)

Known: $\angle \phi$ =   62°35′

V_T =   48V (hypotenuse)

We use $\sin \phi$ = $\dfrac{V_L}{V_T}$.

Step 3. $\sin 62°35′$ = $\dfrac{V_L}{48}$

$\sin 62.58°$ = $\dfrac{V_L}{48}$

0.8877 = $\dfrac{V_L}{48}$

0.8877×48 = $\dfrac{V_L}{48} \times 48$

V_L =   43 V

Step 4.   Hence, the voltage across the inductance (V_L) is 43 V.

7.   Step 1.

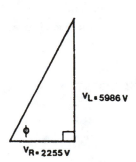

$V_L = 5986$ V

$V_R = 2255$ V

Step 2.   Unknown: $\angle \phi$ =   ?

Known: V_L =   5986 V (side opposite $\angle \phi$)

V_R =   2255 V (side adjacent to $\angle \phi$)

$\therefore \tan \phi$ = $\dfrac{V_L}{V_R}$.

Step 3. $\tan \phi$ = $\dfrac{5986}{2255}$

=   2.6545

$\angle \phi$ =   inv tan (2.6545) (or $\tan^{-1}$, arc tan)

=   69.3577°

=   69°21′

Step 4.   Hence, the phase angle is 69°21′.

8.   Step 1.

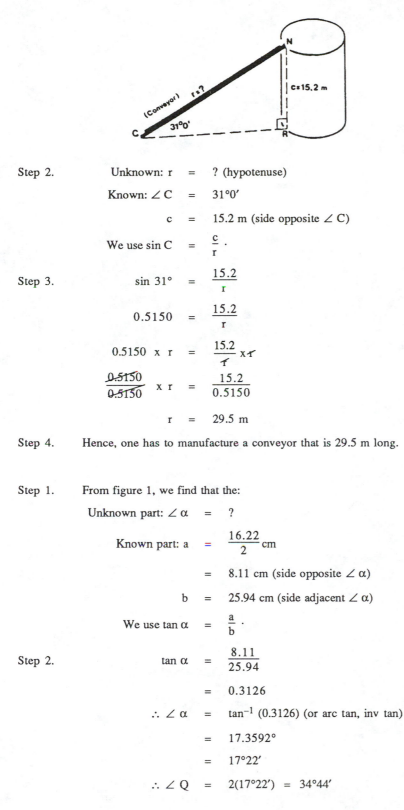

Step 2.   Unknown: r   =   ? (hypotenuse)

Known: $\angle$ C   =   31°0'

c   =   15.2 m (side opposite $\angle$ C)

We use sin C   = $\dfrac{c}{r}$.

Step 3.   sin 31°   = $\dfrac{15.2}{r}$

0.5150   = $\dfrac{15.2}{r}$

0.5150 x r   = $\dfrac{15.2}{r}$ x r

$\dfrac{0.5150}{0.5150}$ x r   = $\dfrac{15.2}{0.5150}$

r   =   29.5 m

Step 4.   Hence, one has to manufacture a conveyor that is 29.5 m long.

9.   Step 1.   From figure 1, we find that the:

Unknown part: $\angle \alpha$ =   ?

Known part: a   = $\dfrac{16.22}{2}$ cm

=   8.11 cm (side opposite $\angle \alpha$)

b   =   25.94 cm (side adjacent $\angle \alpha$)

We use tan α = $\dfrac{a}{b}$.

Step 2.   tan α = $\dfrac{8.11}{25.94}$

=   0.3126

$\therefore \angle \alpha$ = $\tan^{-1}$ (0.3126) (or arc tan, inv tan)

=   17.3592°

=   17°22'

$\therefore \angle$ Q   =   2(17°22') = 34°44'

Step 3.   Hence, the taper angle Q = 34°40' (Rounded to the nearest multiple of 10'.)

Answers to Assignment

1. 14°   3. 15 V   5. 16 V   7. 16°

9. 56 mm   11. 2°0′   13. 6700 m   15. 21.4 m

17. 64°5′   19. 112 mm   21. 34°40′

Part **2**

Module Self-Tests
with Review Analysis

Self Test

Module 1: Whole Numbers and Fractions

Instructions:

1. Write the test without a calculator.
2. Mark your test by using the appropriate answer key. Give yourself 1 mark for a correct answer and 0 for an incorrect answer.
3. Add the marks for each part and enter the sum into Table 1 below.
4. Circle the parts in Table 1 where the "Number Correct" *is less* than the "Number to Pass".
5. Use Table 2 below to review thoroughly the circled parts.

   Example: If you circled Part F, Table 2 shows that you should review Unit 5, Frames *19 to 36*.

Table 1

Part	A	B	C	D	E	F
No. to Pass	10	3	4	8	8	8
No. Correct						

Table 2. Review Analysis

Part A	Unit 2, Frames 1-12
B	Unit 2, Frames 13-25
C	Unit 2, Frames 26-40
D	Unit 3, Frames 33-46; Unit 4
E	Unit 5, Frames 1-18
F	Unit 5, Frames 19-36

Module 1/Self Test

Part A: Indicate the base and the exponent for each of the following:

1. 3^5

2. 5^1

3. 1^5

Change each of the following to exponential notation:

4. 2 x 2 x 2 x 2 =

5. 5 x 5 x 5 =

6. 7 x 7 x 7 x 7 x 7 x 7 x 7 =

1. base = _____

   exponent = _____

2. base = _____

   exponent = _____

3. base = _____

   exponent = _____

4. _____

5. _____

6. _____

Evaluate the following:

7. $2^5 =$ 7. _____

8. $3^3 =$ 8. _____

9. $7^2 =$ 9. _____

10. $\sqrt{64} =$ 10. _____

11. $\sqrt{49} =$ 11. _____

12. $\sqrt{121} =$ 12. _____

Part B: Give the prime factorization for the following.
Put your answer in *exponential form*.

1. 546

1. _____

2. 795

2. _____

Find the L.C.M. of the following, using exponents.

3. 72 and 81

3. _____

4. 14, 21 and 35

4. _____

Part C: Evaluate each of the following using order of operation rules.

1. $16 \div 8 + 2 \cdot 3 =$

1. _____

2. $12 - 2^3 \div 4 \cdot 3 - 3^3 \div 9 =$

2. _____

3. $12 - 2 \cdot (8 - 4 \div 2) =$

4. $[4 + (9 - 3) \cdot 2^3] - 6 =$

5. $34 - 3^3 + 7 =$

6. $26 - 10 + 2 \cdot [7 - (3 - 1)] =$

3.   _____

4.   _____

5.   _____

6.   _____

Part D: Perform the indicated operations and *reduce* the results *to lowest terms*.

1. $1\dfrac{15}{16} + 3\dfrac{13}{16} =$

2. $\dfrac{5}{6} - \dfrac{5}{9} =$

3. $\dfrac{7}{10} + 1\dfrac{3}{4} + \dfrac{1}{2} =$

4. $11\dfrac{3}{10} - 3\dfrac{5}{12} =$

5. $7 + \dfrac{5}{9} =$

6. $3\dfrac{1}{2} - \dfrac{3}{4} =$

7. $4 - 2\dfrac{5}{6} =$

8. $3\dfrac{7}{16} - 2\dfrac{3}{8} =$

1.   _____

2.   _____

3.   _____

4.   _____

5.   _____

6.   _____

7.   _____

8.   _____

9. $7\frac{3}{4} + 4\frac{8}{9} =$ 9. _____

10. $3\frac{1}{3} - 1\frac{1}{4} + 4\frac{7}{8} =$ 10. _____

Part E: Perform the indicated operations and *reduce* results *to lowest terms*.

1. $3 \times \frac{4}{7} =$ 1. _____

2. $\frac{3}{19} \times \frac{5}{6} =$ 2. _____

3. $\left(\frac{3}{4}\right)^3 =$ 3. _____

4. $5\frac{1}{2} \times 2\frac{1}{5} =$ 4. _____

5. $\frac{1}{4} \times 3\frac{1}{2} \times 7 =$ 5. _____

6. $2\frac{2}{5} \times 3\frac{1}{8} \times 1\frac{3}{4} =$ 6. _____

7. $\sqrt{\frac{36}{64}} =$ 7. _____

8. $\frac{3}{4} \times \frac{3}{4} =$ 8. _____

9. $\frac{2}{3} \times 4 =$ 9. _____

10. $\sqrt{\frac{1}{81}}$ 10. _____

Part F: Perform the indicated operations and *reduce* the results *to lowest terms*.

1. $\dfrac{\frac{5}{9}}{\frac{4}{5}} =$ 1. _____

2. $\frac{1}{9} \div \frac{1}{6} =$ 2. _____

3. $\frac{7}{5} \div 14 =$ 3. _____

4. $8 \div \frac{2}{9} =$ 4. _____

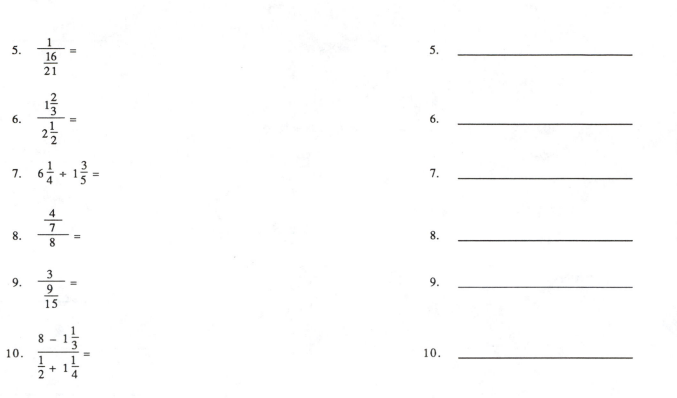

5. $\dfrac{\dfrac{1}{16}}{21} =$ 5. _____

6. $\dfrac{1\frac{2}{3}}{2\frac{1}{2}} =$ 6. _____

7. $6\frac{1}{4} \div 1\frac{3}{5} =$ 7. _____

8. $\dfrac{\dfrac{4}{7}}{8} =$ 8. _____

9. $\dfrac{3}{\dfrac{9}{15}} =$ 9. _____

10. $\dfrac{8 - 1\frac{1}{3}}{\frac{1}{2} + 1\frac{1}{4}} =$ 10. _____

Self Test Answer Key

Module 1

Part A

1. base = 3
    exponent = 5
2. base = 5
    exponent = 1
3. base = 1
    exponent = 5

4. 2^4
5. 5^3
6. 7^7
7.   32
8.   27

9.   49
10. 8
11. 7
12. 11

Part B

1. $2 \cdot 3 \cdot 7 \cdot 13$
2. $3 \cdot 5 \cdot 53$
3.   648
4.   210

Part C

1.   8
2.   3
3.   0
4.   46
5.   14
6.   1

Part D	1. $5\frac{3}{4}$		*Part E*	1. $1\frac{5}{7}$		Part F	1. $\frac{25}{36}$
	2. $\frac{5}{18}$			2. $\frac{5}{38}$			2. $\frac{2}{3}$
	3. $2\frac{19}{20}$			3. $\frac{27}{64}$			3. $\frac{1}{10}$
	4. $7\frac{53}{60}$			4. $12\frac{1}{10}$			4. 36
	5. $7\frac{5}{9}$			5. $6\frac{1}{8}$			5. $1\frac{5}{16}$
	6. $2\frac{3}{4}$			6. $13\frac{1}{8}$			6. $\frac{2}{3}$
	7. $1\frac{1}{6}$			7. $\frac{3}{4}$			7. $3\frac{29}{32}$
	8. $1\frac{1}{16}$			8. $\frac{9}{16}$			8. $\frac{1}{14}$
	9. $12\frac{23}{36}$			9. $2\frac{2}{3}$			9. 5
	10. $6\frac{23}{24}$			10. $\frac{1}{9}$			10. $3\frac{17}{21}$

Self Test

Module 2a: Decimals and the Metric System

Instructions:

1. Write the test without a calculator.
2. Mark your test by using the appropriate answer key. Give yourself 1 mark for a correct answer and 0 for an incorrect answer.
3. Add the marks for each part and enter the sum into Table 1 below.
4. Circle the parts in Table 1 where the "Number Correct" *is less* than the "Number to Pass".
5. Use Table 2 below to review thoroughly the circled parts.

    Example: If you circled Part E, Table 2 shows that you should review Unit 8, Frames *1-21*.

Table 1

Part	A	B	C	D	E	F
No. to Pass	5	5	7	5	5	6
No. Correct						

Table 2. Review Analysis

Part A	Unit 6, Frames 1-16
B	Unit 6, Frames 17-25
C	Unit 6, Frames 26-38
D	Unit 7
E	Unit 8, Frames 1-21
F	Unit 8, Frames 22-39

Module 2a/Self Test

Part A: Perform the indicated operations:

1. $3.5 + 0.2 + 12 =$ 1. _____

2. $2 - 0.0075 =$ 2. _____

3. $126.025 + 0.0014 + 140 =$ 3. _____

4. $0.95 - 0.087 =$ 4. _____

5. $600 - 0.892 =$ 5. _____

Part B: Perform the indicated operations:

1.   4.97 x 1.4 =   1. _____

2. $(0.2)^4$ =   2. _____

3.   32.08 x 0.012 =   3. _____

4.   0.01 x 1.5 =   4. _____

5.   0.255 x 0.0056 =   5. _____

6.   32 x 1.04 x 0.25 =   6. _____

Part C: Divide:

1.   0.00012 ÷ 0.03 -   1. _____

2.   40 ÷ 12.8 =   2. _____

3.   3.74 ÷ 100 =   3. _____

4.   3.1408 ÷ 3.02 =   4. _____

5.   0.55 ÷ 0.01 =   5. _____

6.   0.0112 ÷ 2.8 =   6. _____

Divide and *round off* the answers to *two decimal* places.

7.   27.18 ÷ 13.6 =   7. _____

8.   2 ÷ 0.34 =   8. _____

Part D: Length

Perform the following conversions.

1.   4500 cm   = _____ m   1. _____

2.   0.0006 km   = _____ m   2. _____

3.   0.074 mm   = _____ cm   3. _____

4.   8500 dam   = _____ km   4. _____

5.   9 hm   = _____ dm   5. _____

6.   0.125 m   = _____ cm   6. _____

Part E: Area and Volume

Perform the following conversions.

1.   79 000 m^2 = _____ km^2 1. _____

2.   0.03 hm^2 = _____ m^2 2. _____

3.   86 cm^2 = _____ mm^2 3. _____

4.   60 000 mm^3 = _____ cm^3 4. _____

5.   2500 cm^3 = _____ dm^3 5. _____

6.   750 dm^3 = _____ m^3 6. _____

Part F: Perform the following conversions.

1.   1.8 kL = _____ L   1. _____

2.   2.37 L = _____ mL   2. _____

3.   0.082 mL = _____ cm^3 3. _____

4.   0.0178 m^3 = _____ L   4. _____

5.   3.9 kg = _____ g   5. _____

6.   0.287 g = _____ mg   6. _____

7.   875 g = _____ kg   7. _____

Self Test Answer Key

Module 2a

Part A		*Part B*		*Part C*	
1.	15.7	1.	6.958	1.	0.004
2.	1.9925	2.	0.0016	2.	3.125
3.	266.0264	3.	0.38496	3.	0.0374
4.	0.863	4.	0.015	4.	1.04
5.	599.108	5.	0.001 428	5.	55
		6.	8.32	6.	0.004
				7.	2.00
				8.	5.88

Part D		*Part E*		*Part F*	
1.	45 m	1.	0.079 km^2	1.	1800 L
2.	0.6 m	2.	300 m^2	2.	2370 mL
3.	0.0074 cm	3.	8600 mm^2	3.	0.082 cm^3
4.	85 km	4.	60 cm^3	4.	17.8 L
5.	9000 dm	5.	2.5 dm^3	5.	3900 g
6.	12.5 cm	6.	0.75 m^3	6.	287 mg
				7.	0.875 kg

Self Test

Module 2b: Percent

Instructions:

1. Write the test without a calculator.
2. Mark your test by using the appropriate answer key. Give yourself 1 mark for a correct answer and 0 for an incorrect answer.
3. Add the marks for each part and enter the sum into Table 1 below.
4. Circle the parts in Table 1 where the "Number Correct" *is less* than the "Number to Pass".
5. Use Table 2 below to review thoroughly the circled parts.

    Example: If you circled Part A, Table 2 shows that you should review Unit 9, Frames *1 to 26*.

Table 1

Part	A	B	C	D	E
No. to Pass	8	8	3	12	2
No. Correct					

Table 2. Review Analysis

Part A	Unit 9, Frames 1-26
B	Unit 9, Frames 27-37
C	Unit 9, Frames 10-16
D	Unit 10, Frames 1-43
E	Unit 10, Frames 44-48

Module 2b/Self Test

Part A: Change the following percents to decimals:

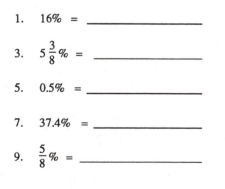
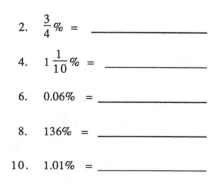

1. 16% = _____

2. $\frac{3}{4}$% = _____

3. $5\frac{3}{8}$% = _____

4. $1\frac{1}{10}$% = _____

5. 0.5% = _____

6. 0.06% = _____

7. 37.4% = _____

8. 136% = _____

9. $\frac{5}{8}$% = _____

10. 1.01% = _____

Part B: Change the following numerals to a percent:

1. $\frac{3}{10}$ = _____ %

2. 0.0042 = _____ %

3. 0.75 = _____ %

4. $\frac{1}{4}$ = _____ %

5. $1\frac{1}{10}$ = _____ %

6. 0.728 = _____ %

7. 0.04 = _____ %

8. $\frac{2}{3}$ = _____ %

9. 1.7 = _____ %

10. 2 = _____ %

Part C: Change each of the following fractions to decimals.
Round to three decimal places where appropriate.

1. $\frac{2}{5}$ =

1. _____

2. $\frac{7}{22}$ =

2. _____

3. $\frac{5}{6}$ =

3. _____

4. $2\frac{3}{8}$ =

4. _____

Part D: Find:

1. 10.5% of 86 = _____

2. 6.3% of what number is 0.63? _____

3. What percent of 48 is 1.2? _____

4. 0.75% of 360 = _____

5. What percent of 16 is 20? _____

6. $\frac{3}{4}$ % of what number is 6? _____

7. 135% of 200 = _____

8. 0.4% of 0.7 = _____

9. 8% of what number is 20? _____

10. What percent of $\frac{2}{3}$ is $\frac{1}{3}$? _____

11. 120% of what number is 72? _____

12. $\frac{3}{4}$ % of 400 = _____

13. What percent of 75 is 30? _____

14. $5\frac{1}{4}$ % of 80 = _____

15. 0.8% of what number is 0.96? _____

Part E:

1. Paper cups are listed at 2 cents each with a discount of 5% when 1000 cups are ordered. What is the amount of the discount on 1000 cups?

   1. $ _____

2. One dealer lists a car at $4250 with a 14% end-of-the-model year discount. Another dealer lists the same car at $3700 with no end-of-the-model year discount. Which is the better buy and by how much?

   2. $ _____

3. A car sells for $9500.00 with all options included. How much of this amount does a dealer receive if she pays her sales people an 8% commission?

   3. $ _____

4. What will the total price be on a car selling for $4155 when the sales tax is 7%?

   4. $ _____

Self Test Answer Key

Module 2b

Part A						*Part B*		
	1.	0.16	6.	0.000 6			1.	30%
	2.	0.007 5	7.	0.374			2.	0.42%
	3.	0.053 75	8.	1.36			3.	75%
	4.	0.011	9.	0.006 25			4.	25%
	5.	0.005	10.	0.010 1			5.	110%
							6.	72.8%
Part C	1.	0.4	3.	0.833			7.	4%
	2.	0.318	4.	2.375			8.	$66\frac{2}{3}$%
							9.	170%
Part D	1.	9.03	6.	800	11.	60	10.	200%
	2.	10	7.	270	12.	3		
	3.	2.5%	8.	0.002 8	13.	40%	*Part E*	1.	$1.00 discount
	4.	2.7	9.	250	14.	4.2		2.	Dealer #1 better by $45.00
	5.	125%	10.	50%	15.	120		3.	Dealer receives $8740.00.
								4.	Total cost is $4445.85.

Self Test

Module 3: Operations with Integers, Exponents

Instructions:

1. Mark your test by using the appropriate answer key. Give yourself 1 mark for a correct answer and 0 for an incorrect answer.
2. Add the marks for each part and enter the sum into Table 1 below.
3. Circle the parts in Table 1 where the "Number Correct" *is less* than the "Number to Pass".
4. Use Table 2 below to review thoroughly the circled parts.

    Example: If you circled Part C, Table 2 shows that you should review Unit 13.

Table 1

Part	A	B	C	D	E	F	G
No. to Pass	7	6	9	6	7	2	2
No. Correct							

Table 2. Review Analysis

Part A	Unit 11
B	Unit 12
C	Unit 13
D	Unit 14, Frames 1-39
E	Unit 14, Frames 40-52
F	Unit 15, Frames 1-9
G	Unit 15, Frames 10-34

Module 3/Self Test

Part A: Answer each of the following 8 questions by indicating which are true and which are false.
MARK A "T" OR "F" IN THE BOX FOLLOWING THE QUESTIONS.

1.	$+10 = 10$		5.	$\lvert -5 \rvert = -5$	
2.	$\lvert +7 \rvert = 7$		6.	$-(-3) = +3$	
3.	$-[-(-12)] = 12$		7.	$-15 > 5$	
4.	$-3 < 0$		8.	$12 > 0$	

Part B: Evaluate each of the following:

1. $(+8) + (-8) =$ 1. _____

2. $(-6) + (-7) + (+8) =$ 2. _____

3. $2 - 3 + 5 - 9 + 10 - 1 =$ 3. _____

4. $(-2) - (-8) - (+4) + (-3) =$ 4. _____

5. $(+6) + (-3) - (-6) - (+7) =$ 5. _____

6. $0 - (-9) =$ 6. _____

7. $(-12) - 0 =$ 7. _____

8. $0 + (-7) =$ 8. _____

Part C: Evaluate each of the following:

1. $6 \cdot (-3) \div (-9) =$ 1. _____

2. $(-9) \div (-3) \div (-1) =$ 2. _____

3. $0 \div 6 \div (-2) \div (-6) =$ 3. _____

4. $(-5)^2 =$ 4. _____

5. $-7^2 =$ 5. _____

6. $(-10)^0 =$ 6. _____

7. $(-1)^{62} =$ 7. _____

8. $(-2)^5 =$ 8. _____

9. $(-2)^3 \cdot 4 + 5(-2)^2 \div (-4) =$ 9. _____

10. $(-7 + 5)^4 + (6 - 4)^5 =$ 10. _____

11. $(-4)^2 + (7 - 9)^2 + 5(-6) =$ 11. _____

12. $-4^2 + (7 - 9)^2 + 5(-6)^0 =$ 12. _____

Part D: Simplify the following by applying one of the properties of exponents:

1. $5a^0$ (where $a \neq 0$) =   1. _____

2. $(a^4)^3$ =   2. _____

3. $(3x^2y^5)^2$ =   3. _____

4. $y^3 \cdot y^6$ =   4. _____

5. $x^{15} \div x^3$ =   5. _____

6. $\left(\dfrac{-1}{a^6}\right)^4$ =   6. _____

7. $(y-3)^5 \div (y-3)^3$ =   7. _____

8. $\left(\dfrac{2x^3y}{a^2b^3}\right)^5$ =   8. _____

9. $(3xy)^4 \, (3xy)^2$ =   9. _____

Part E: Evaluate each of the following:

1. 2^{-5} =   1. _____

2. $\left(\dfrac{3}{4}\right)^{-2}$ 2. _____

Simplify the following so that no final answer has negative exponents. Reduce terms where appropriate.

3. $\dfrac{a^{-3}}{3b}$ =   3. _____

4. $\dfrac{2x}{6y^{-5}}$ =   4. _____

5. $\dfrac{5x^{-4}}{3y^{-3}}$ =   5. _____

6. $\left(\dfrac{-2p^{-3}q^2}{5m^{-4}n}\right)^{-3}$ =   6. _____

7. $5(a+b)^{-6}$ =   7. _____

8. $a^{-2} \cdot a^{-6}$ =   8. _____

9. $2^3q^0p^{-3}s^4$ =   9. _____

10. $\dfrac{10^{-6} \cdot 10^{13}}{10^4 \cdot 10^{-2}}$ =   10. _____

Part F:

1. Evaluate $(9y^{-10})(2y^8)$ when $y = 3$.

   1. _____

2. Evaluate $(a^3b^{-5}c^2)^0(a^3b^2c)^2$ when $a = 1$, $b = 2$, $c = 5$.

   2. _____

3. Evaluate $\left(\dfrac{2m^3n^{-1}}{t^{-2}}\right)^{-2}$ when $m = 1$, $n = 4$, $t = 2$.

   3. _____

Part G: Use scientific notation to perform the indicated operations. Leave your answer in scientific notation.

1. $0.000\ 000\ 018 \times 200\ 000\ 000$

   1. _____

2. $4\ 500\ 000 \div 0.000\ 15$

   2. _____

3. $\dfrac{0.000\ 36 \times 500\ 000\ 000}{18\ 000\ 000 \times 0.000\ 025}$

   3. _____

Self Test Answer Key

Module 3

Part A			*Part B*			*Part C*		
	1.	True		1.	0		1.	2
	2.	True		2.	−5		2.	−3
	3.	False		3.	4		3.	−6
	4.	True		4.	−1		4.	25
	5.	False		5.	2		5.	−49
	6.	True		6.	9		6.	1
	7.	False		7.	−12		7.	1
	8.	True		8.	−7		8.	−32
							9.	−37
							10.	48
							11.	−10
							12.	−7

Part D   1. 5

2. a^{12}

3. $9x^4y^{10}$

4. y^9

5. x^{12}

6. $\dfrac{1}{a^{24}}$

7. $(y - 3)^2$

8. $\dfrac{32x^{15}y^5}{a^{10}b^{15}}$

9. $(3xy)^6$ or $729x^6y^6$

Part E   1. $\dfrac{1}{32}$

2. $1\dfrac{7}{9}$

3. $\dfrac{1}{3a^3b}$

4. $\dfrac{xy^5}{3}$

5. $\dfrac{5y^3}{3x^4}$

6. $-\dfrac{125p^9n^3}{8q^6m^{12}}$

7. $\dfrac{5}{(a + b)^6}$

8. $\dfrac{1}{a^8}$

9. $\dfrac{8s^4}{p^3}$

10. 10^5 or 100 000

Part F   1. 2

2. 400

3. $\dfrac{1}{4}$

Part G   1. 3.6×10^0

2. 3.0×10^{10}

3. 4.0×10^2

Self Test

Module 4: Linear Equations, Operations with Algebraic Expressions

Instructions:

1. Mark your test by using the appropriate answer key. Give yourself 1 mark for a correct answer and 0 for an incorrect answer.
2. Add the marks for each part and enter the sum into Table 1 below.
3. Circle the parts in Table 1 where the "Number Correct" *is less* than the "Number to Pass".
4. Use Table 2 below to review thoroughly the circled parts.
    Example: If you circled Part D, Table 2 shows that you should review Unit 19, Frames *1 – 17*.

Table 1

Part	A	B	C	D	E	F	G
No. to Pass	4	4	2	4	5	3	5
No. Correct							

Table 2. Review Analysis

Part A	Unit 16, Frames 1-30
B	Unit 17
C	Unit 18
D	Unit 19, Frames 1-17
E	Unit 20
F	Unit 16, Frames 31-36
G	Unit 19, Frames 18-36

Module 4/Self Test

Part A: Simplify the following expressions by combining like terms.

1. $9a - 2a =$

    1. _____

2. $9x^2 + 8xy - 7x^2 + xy =$

    2. _____

3. $6r^2s + 4rs^2 - rs^2 - 3r^2s =$

    3. _____

4. $-(3m - 2n) + 7m - n =$

    4. _____

5. $(x - 5) - (3x + 7) + (2x + 9) =$

    5. _____

6. $2y^2 + 8y - (y^2 + 5y - 9) =$

    6. _____

Part B: Solve each of the following equations:

1. $12x = 96$ 1.   x = _____

2. $6m - 7 = 23$ 2.   m = _____

3. $5 - 3s = 25 - 7s$ 3.   s = _____

4. $8(x - 4) = 2(x + 2)$ 4.   x = _____

5. $15 - (3a - 10) = a + 21 - (2a + 4)$ 5.   a = _____

6. $-3(m - 1) + m = 2(m - 1) - 3$ 6.   m = _____

Part C: Solve the following word problems. (Show all steps)

1.   The sum of 2 numbers is 76. One of the numbers is three times the other. Find the numbers.

1.   _____

2.   The length of a rectangle is 5 cm less than 4 times its width, and its perimeter is 90 cm. Find the length of the rectangle.

2.   Length = _____

3.   A wire 54 cm long is cut into three pieces. The second piece is twice the length of the first, and the third piece is three times the length of the first. Find the length of each piece.

3.   _____

4.   A developer bought 900 acres of land for $680 000. Some was bought at $600 per acre, some for $800 per acre and some $1000 per acre. How much of each did he buy if he bought twice as many acres at $600 as he did at $1000?

4.   _____

Part D: Multiply the following as indicated.

1. $(6m^5)(5m^3) =$ 1. _____

2. $(-a^2b^{-4})(3a^2b^4) =$ 2. _____

3. $4(8x - 12y) =$ 3. _____

4. $-3a(2a - 3b + 2) =$ 4. _____

5. $2mn^3(4m + mn - 6) =$ 5. _____

Part E: Divide the following.

1. $(-10x^3yz^2) \div (5x^2y^3z) =$ 1. _____

2. $\dfrac{4b^2 - 2b^3}{2b} =$ 2. _____

3. $(6xy^2 - 12xy^3) \div (-6xy) =$ 3. _____

4. $\dfrac{42ax - 28a + 70ay}{-14a} =$ 4. _____

Factor the greatest common monomial from each of the following expressions:

5. $3x - 9y =$ 5. _____

6. $16m^2n - 4m =$ 6. _____

7. $3a^3b - a^2b + 5ab^2 =$ 7. _____

Part F: Simplify the following expressions by removing the symbols of grouping and collecting like terms.

1. $(-a^2 + ab + b^2) - (-4a^2 - ab + b^2) - 2ab =$

1. _____

2. $2a - [3b - (c + 4)] =$

2. _____

3. $\{z - [3w - (2w + 4z)]\} + 3w =$

3. _____

4. $6x - \{2y - [-2x - (x - y) - 3x]\} =$

4. _____

5. $[5x - (4x + 2y)] - [8 - (7y + 3)] =$

5. _____

Part G: Multiply vertically:

1. $(x^2 + 5x - 3)(x - 4) =$

1. _____

2. $(2y^2 - 4y + 3)(4y^2 + y - 5) =$

2. _____

Multiply, using any *shortcut* method which applies:

3. $(5w - 1)(5w + 1) =$

3. _____

4. $(2a + 7)(3a + 1) =$

4. _____

5. $(b + 4)^2$ =

5.   _____

6. $(4a - 5) (4a + 5)$ =

6.   _____

7. $(x - 5y) (5y - x)$ =

7.   _____

Self Test Answer Key

Module 4

Part A			*Part B*			*Part C*		
	1.	$7a$		1.	$x = 8$		1.	19 and 57
	2.	$2x^2 + 9xy$		2.	$m = 5$		2.	35 cm
	3.	$3r^2s + 3rs^2$		3.	$s = 5$		3.	1st piece = 9 cm
	4.	$4m + n$		4.	$x = 6$			2nd piece = 18cm
	5.	-3		5.	$a = 4$			3rd piece = 27 cm
	6.	$y^2 + 3y + 9$		6.	$m = 2$		4.	400 acres @ $600
								300 acres @ $800
								200 acres @ $1000

Part D			*Part E*			*Part F*		
	1.	$30m^8$		1.	$\dfrac{-2xz}{y2}$		1.	$3a^2$
	2.	$-3a^4$		2.	$2b - b^2$		2.	$2a - 3b + c + 4$
	3.	$32x - 48y$		3.	$-y + 2y^2$		3.	$2w + 5z$
	4.	$-6a^2 + 9ab - 6a$		4.	$-3x + 2 - 5y$		4.	$-y$
	5.	$8m^2n^3 + 2m^2n^4 - 12mn^3$		5.	$3 (x - 3y)$		5.	$x + 5y - 5$
				6.	$4m (4mn - 1)$			
				7.	$ab (3a^2 - a + 5b)$			

Part G					
	1.	$x^3 + x^2 - 23x + 12$		5.	$b^2 + 8b + 16$
	2.	$8y^4 - 14y^3 - 2y^2 + 23y - 15$		6.	$16a^2 - 25$
	3.	$25w^2 - 1$		7.	$-x^2 + 10xy - 25y^2$
	4.	$6a^2 + 23a + 7$			

Self Test

Module 5a: Operations with Fractional Expressions, Solving Formulas

Instructions:

1. Mark your test by using the appropriate answer key. Give yourself 1 mark for a correct answer and 0 for an incorrect answer.

2. Add the marks for each part and enter the sum in Table 1 below.

3. Circle the parts in Table 1 where the "Number Correct" *is less* than the "Number to Pass".

4. Use Table 2 below to review thoroughly the circled parts.

   Example: If you circled Part A, Table 2 shows that you should review Unit 21 and Unit 22.

Table 1

Part	A	B	C
No. to Pass	5	4	3
No. Correct			

Table 2. Review Analysis

Part A	Unit 21, Unit 22
B	Unit 23
C	Unit 24

Module 5a/Self Test

Part A: *Multiply or divide* as indicated and reduce to lowest terms where appropriate:

1. $\dfrac{8a^3}{3b^2} \cdot \dfrac{5b}{32a} =$

   1. _____

2. $\dfrac{9(x-4)}{8y^2} \div \dfrac{27(x-4)}{12y} =$

   2. _____

3. $\left(\dfrac{-8m}{3pq}\right)\left(\dfrac{-6p^2}{2q}\right) =$

   3. _____

4. $10x^3 \cdot \dfrac{2y^2}{-5x^2z} =$

   4. _____

5. $\dfrac{3m}{5m^2 - 10m} \cdot \dfrac{3m-6}{n^2} =$

   5. _____

6. $\dfrac{\dfrac{3V}{4(R-1)}}{\dfrac{V^2}{6R^3}} =$

   6. _____

7. $\dfrac{4(a-1)}{-3b} \div (2a) =$

7. _____

Part B: Combine the following and reduce to lowest terms where appropriate.

1. $\dfrac{10x}{3z} - \dfrac{4x}{3z} =$

1. _____

2. $\dfrac{5}{2a} - \dfrac{3}{4a^2} + \dfrac{1}{a} =$

2. _____

3. $\dfrac{9a}{x} - \dfrac{5a}{y} =$

3. _____

4. $\dfrac{3p-1}{5q} - \dfrac{3p+4}{5q} =$

4. _____

5. $\dfrac{2}{m} - \dfrac{1}{n} + \dfrac{a}{p} =$

5. _____

6. $\dfrac{V}{18R^4Z^2} - \dfrac{P}{27RZ^3} =$

6. _____

Part C: Solve each of the following equations and *check* your solution.

1. $\dfrac{1}{5}q = 12$

1. q = _____

2. $\dfrac{1}{6}m + \dfrac{5}{6} = \dfrac{2}{3}$

2. m = _____

3. $\dfrac{x}{2} - \dfrac{2x+1}{5} = \dfrac{3}{4}$

3. x = _____

4. $\dfrac{2}{R} - \dfrac{3}{7} = -1$

4. R = _____

5. $\dfrac{1}{3d-4} = \dfrac{5}{4}$

5. d = _____

Self Test Answer Key

Module 5a

Part A   1. $\dfrac{a^2}{2b}$

2. $\dfrac{1}{2y}$

3. $\dfrac{8mp}{q^2}$

4. $\dfrac{-4xy^2}{z}$

5. $\dfrac{9}{5n^2}$

6. $\dfrac{9R^3}{2V(R-1)}$

7. $\dfrac{2(a-1)}{-3ab}$

Part B   1. $\dfrac{2x}{z}$

2. $\dfrac{14a-3}{4a^2}$

3. $\dfrac{9ay-5ax}{xy}$

4. $-\dfrac{1}{q}$

5. $\dfrac{2np-mp+amn}{mnp}$

6. $\dfrac{3ZV-2PR^3}{54R^4Z^3}$

Part C   1. 60

2. -1

3. $\dfrac{19}{2}$

4. $-\dfrac{7}{2}$

5. $\dfrac{8}{5}$

Self Test

Module 5b: Operations with Fractional Expressions, Solving Formulas

Instructions:

1. Mark your test by using the appropriate answer key. Give yourself 1 mark for a correct answer and 0 for an incorrect answer.

2. Add the marks for each part and enter the sum in Table 1 below.

3. Circle the parts in Table 1 where the "Number Correct" *is less* than the "Number to Pass".

4. Use Table 2 below to review thoroughly the circled parts.

   Example: If you circled Part E, Table 2 shows that you should review Unit 27.

Table 1

Part	D	E	F
No. to Pass	4	3	3
No. Correct			

Table 2. Review Analysis

Part D	Unit 25
E	Unit 27
F	Unit 26

Module 5b/Self Test

Part D:

1. 24 kg of an alloy contains 4 kg of silver, 8 kg of copper and 12 kg of lead. Find:

   (a)   the ratio of silver to copper: _____

   (b)   the ratio of copper to lead: _____

   (c)   the ratio of silver to the total amount of alloy: _____

2. Change the following to a fractional equation and solve for the unknown.

   $6 : x :: 8 : 3$

   2. $x =$ _____

Given: 1 in   = 2.54 cm   1 quart = 2 pints

                1 quart = 1.14 L   1 ft   = 12 in

                1 kg   = 2.20 lb   1 lb   = 16 oz

                C   = $\frac{5}{9}$ (F − 32)   F   = $\frac{9}{5}$ C + 32

Perform the following conversions, using conversion ratios. Round your answers appropriately, if necessary.

3.   49 oz = _____ kg

                                                                    3.   _____

4.   35 cm = _____ ft

                                                                    4.   _____

5.   12.5 pints = _____ L

                                                                    5.   _____

6.   −25°C = _____ °F

                                                                    6.   _____

7.   95°F = _____ °C

                                                                    7.   _____

Part E: Solve the following formulas for the indicated variable.

   1.   If S = m + ut, solve for u.

                                                                    1.   u = _____

2. If $f = h - \dfrac{c}{n}$, solve for c.

2. c = _____

3. If $C = 2\pi r$, solve for r.

3. r = _____

4. If $C = \dfrac{5}{9}(F - 32)$, solve for F.

4. F = _____

5. If $P = \dfrac{S}{1 + rt}$, solve for r.

5. r = _____

Part F: Solve the following word problems. *Show all steps.*

1. If $\dfrac{1}{5}$ of a number is added to the number, the result is 96.

    What is the number?

1. Number is: _____

2. The perimeter of an isosceles triangle (two sides are equal) is 60 cm. If the unequal side is 5 cm more than half the length of one of the equal sides, what is the length of the unequal side?

2. Length: _____

3.   9 g of calcium will combine with 32 g of chlorine. How much chlorine will combine with 10 g of calcium?

3. _____

4.   How many litres of water must be added to 10 litres of a 40% alcohol in water solution to decrease it to a 15% alcohol solution?

4. _____

5.   A credit union invests $100 000. Part of the investment has a return of 11% and the other part earns 16% interest. If the interest earned totals $13 000 per year, find out how much the credit union invests at 11% and how much at 16%.

5. _____

Self Test Answer Key

Module 5b

Part D   1. (a) 1:2

  (b) 2:3

  (c) 1:6

2. $x = \dfrac{9}{4}$

3. 1.4 kg

4. 1.1 ft

5. 7.125 L

6. −13°F

7. 35°C

Part E   1. $u = \dfrac{s - m}{t}$

2. $c = n(h - f)$

3. $r = \dfrac{c}{2\pi}$

4. $F = \dfrac{9}{5}C + 32$

5. $r = \dfrac{S - P}{Pt}$

Part F   1. 80

2. 16 cm

3. $35.\overline{5}$ g

4. $16\dfrac{2}{3}$ L

5. $60 000.00 @ 11%
     $40 000.00 @ 16%

Self Test

Module 6: Functions, Graphs, Systems of Linear Equations

Instructions:

1.   Mark your test by using the appropriate answer key. Give yourself 1 mark for a correct answer and 0 for an incorrect answer.

2.   Add the marks for each part and enter the sum in Table 1 below.

3.   Circle the parts in Table 1 where the "Number Correct" *is less* than the "Number to Pass".

4.   Use Table 2 below to review thoroughly the circled parts.

   Example: If you circled Part E, Table 2 shows that you should review Unit 30, *Frames 31 to 35*.

Table 1

Part	A	B	C	D	E
No. to Pass	8	3	3	3	1
No. Correct					

Table 2. Review Analysis

Part A	Unit 28, Frames 1-17, 26-29
B	Unit 28, Frames 18-25
C	Unit 29
D	Unit 30, Frames 1-30
E	Unit 30, Frames 31-35

Module 6/Self Test

Part A: Locate the following points on the graph below, by letter:

1.   A(3, 2)

2.   B(0, 2)

3.   C(−3, 0)

4.   D(−1, 4)

5.   E(3, −4)

Name the points located on the graph below, using ordered pair

6. A(   )

7. B(   )

8. C(   )

9. D(   )

10. E(   )

Part B: Make a table of values *and graph the following*, using the graph sheets provided.
Each graph must have the *X and Y axis clearly labelled, and a scale for each axis must be shown.*

1.   y = x – 3

x			
y			

2.   y = 3x – 2

x			
y			

3. $y = \dfrac{1}{3} x - 2$

x			
y			

4. $y = 3 - x^2$

x			
y			

Part C: Graph the following linear equations and then determine the quantities indicated.

1. $y = x + 1$

(a) x-intercept   = _____

(b) y-intercept   = _____

(c) slope   = _____

2. y = 8 − 2x
    (a) x-intercept   = _____
    (b) y-intercept   = _____
    (c) slope   = _____

3. 2x + 3y = 6
    (a) x-intercept   = _____
    (b) y-intercept   = _____
    (c) slope   = _____

4. 3y = 12 − 2x
    (a) x-intercept   = _____
    (b) y-intercept   = _____
    (c) slope   = _____

5. y − x = 4
    (a) x-intercept   = _____
    (b) y-intercept   = _____
    (c) slope   = _____

Part D: Solve the following systems of linear equations using any method you wish.

1. $3x + y = 18$
      $x - y = -2$

x = _____   y = _____

2. $2x - 3y = 6$
      $2x + 3y = -18$

x = _____   y = _____

3. $5x - 2y = 3$
      $3x - 4y = -1$

x = _____   y = _____

4. $2a - b = 2$
      $4a - 2b = 4$

a = _____   b = _____

5. $m = 5n + 2$
      $15n - 2m = 1$

m = _____   n = _____

Part E:

1. The sum of two numbers is 31. If twice the larger number is subtracted from four times the smaller number, the result is 10. What are the numbers?

1. _____

2. A rectangular field has a perimeter of 160 metres. If the width is 8 metres smaller than the length, what are the dimensions of the fields? Perimeter = 2 (length + width)

2. _____

Self Test Answer Key

Module 6

Part A   1 to 5

6. (2, 4)

7. (−3, 5)

8. (−4, −1)

9. (0, −3)

10. (6, −5)

Part B

1.

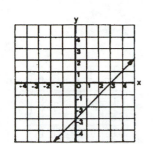

2.

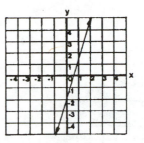

3.

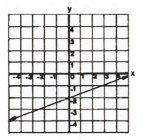

4.

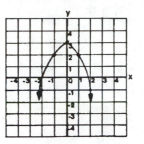

Part C

1.   (a)   x-intercept = −1

    (b)   y-intercept = 1

    (c)   slope = 1

2.   (a)   x-intercept = 4

    (b)   y-intercept = 8

    (c)   slope = −2

3.   (a)   x-intercept = 3

    (b)   y-intercept = 2

    (c)   slope = $-\dfrac{2}{3}$

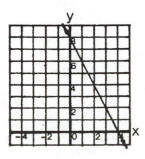

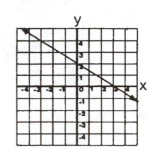

4.   (a)   x-intercept = 6

      (b)   y-intercept = 4

      (c)   slope = $-\dfrac{2}{3}$

5.   (a)   x-intercept = –4

      (b)   y-intercept = 4

      (c)   slope = 1

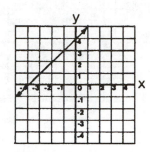

Part D   1.   x = 4   y = 6

            2.   x = –3   y = –4

            3.   x = 1   y = 1

4.   no unique solution

5.   m = 7   n = 1

Part E   1.   12 and 19

            2.   36 m by 44 m

Self Test

Module 7: Basic Geometry and Trigonometry

Instructions:

1. Mark your test by using the appropriate answer key. Give yourself 1 mark for a correct answer and 0 for an incorrect answer.

2. Add the marks for each part and enter the sum in Table 1 below.

3. Circle the parts in Table 1 where the "Number Correct" *is less* than the "Number to Pass".

4. Use Table 2 below to review thoroughly the circled parts.

   Example: If you circled Part E, Table 2 shows that you should review Unit 33.

Table 1

Part	A	B	C	D	E	F
No. to Pass	4	2	1	6	8	2
No. Correct						

Table 2. Review Analysis

Part A	Unit 32, Frames 11-16
B	Unit 32, Frames 17-21
C	*Unit 34*, Frames 1-15; *Unit 35*, Frames 1-8
D	Unit 34, Frames 16-34
E	Unit 33
F	Unit 36

Part A: Find the measure of the unknown angle or angles.

1.

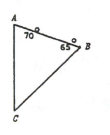

2.

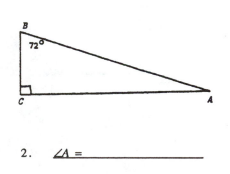

1.   ∠C = _____

2.   ∠A = _____

3.

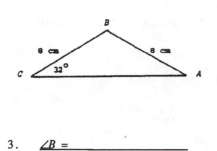

4.

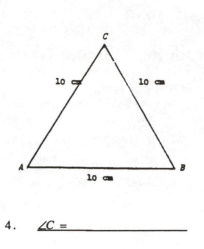

3.   ∠B = _____

4.   ∠C = _____

Part B: Use the pythagorean theorem to find the length of the unknown side in the following triangles:

1.

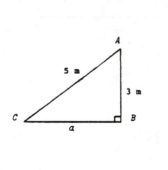

2.

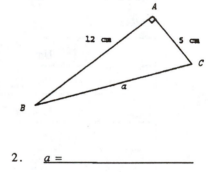

1.   a = _____

2.   a = _____

Part C: Answer the following questions correctly.

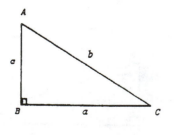

Sin A = _____

Cot C = _____

Cos A = _____

Sec C = _____

Csc A = _____

Part D: Find the indicated quantity.
Give angles to the nearest minute, and lengths to four significant digits.

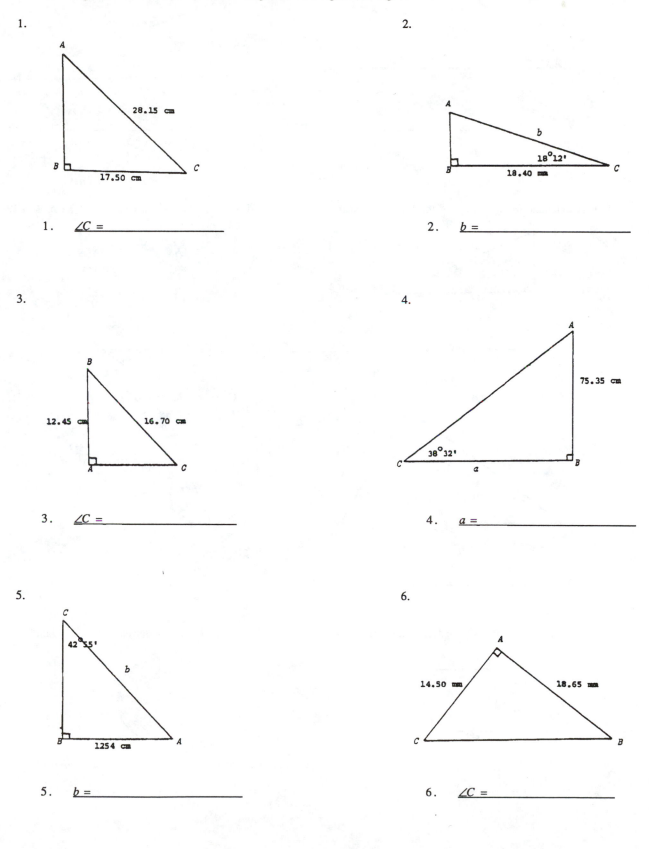

1.

28.15 cm

17.50 cm

1.   ∠C = _____

2.

18°12'

18.40 mm

2.   b = _____

3.

12.45 cm   16.70 cm

3.   ∠C = _____

4.

75.35 cm

38°32'

a

4.   a = _____

5.

42°55'

b

1254 cm

5.   b = _____

6.

14.50 mm   18.65 mm

6.   ∠C = _____

7.

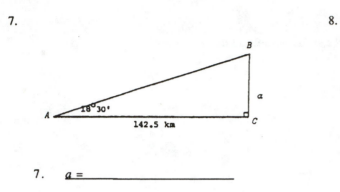

8.

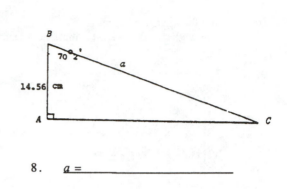

7. $a =$ _____

8. $a =$ _____

Part E:

1.   Calculate *the perimeter* of the following triangle.

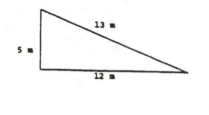

1.   _____

2.   Find *the circumference* of a circle with a radius of 12 cm.

2.   _____

3.   Calculate *the area* of the following triangle:

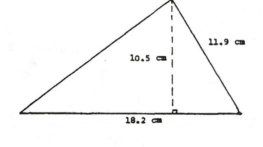

3.   _____

4.   Calculate *the area* of the following trapezoid:

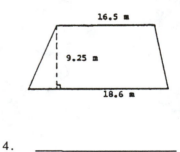

4.   _____

5.   Calculate *the area* of the following parallelogram:

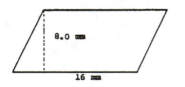

5.   _____

6.   Find *the area of the shaded part* of the diagram below:

6.   _____

7. Find *the volume of the rectangular solid* illustrated below:

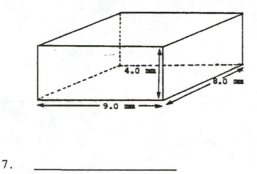

7. _____

8. Calculate *the volume* of the cone illustrated below:

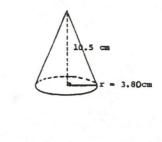

8. _____

9. Calculate *the surface area* of a sphere with a radius of 4.00 m.

9. _____

10. On the cylinder illustrated below, r = 25.0 cm and h = 1.05 m. Calculate *the volume* of this cylinder.

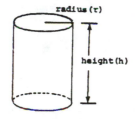

10. _____

Part F:

1. A lighthouse is 150 m high. The angle of elevation to the top of the lighthouse from a point on shore is 25° 40'. Find the distance to the base of the lighthouse from the point on shore.

1. _____

2.   A railroad track rises 1.100 m for every 50.00 m measured *along the track*. Find the angle of inclination of the track.

2.   _____

3.   The angle of depression of an aircraft carrier from an approaching airplane is 48° 22´. If the plane is flying at an altitude of 1240 m, find the horizontal distance to the carrier.

3.   _____

4.   In the isosceles triangle below, find the altitude "*a*".

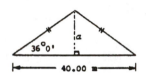

4.   _____

Self Test Answer Key

Module 7

Part A	1. 45°	2. 18°	3. 116°	4. 60°

Part B   1. 4 m   2. 13 cm

Part C $\sin A = \dfrac{a}{b}$ $\cot C = \dfrac{a}{c}$ $\cos A = \dfrac{c}{b}$ $\sec C = \dfrac{b}{a}$ $\csc A = \dfrac{b}{a}$

Note: For the remaining questions, if your answers are not rounded to the proper number of significant digits, review Unit 30 in textbook.

Part D   1. 51°34´   5. 1842 cm   *Part E*   1. 30 m   6. 6.8 m^2

2. 19.37 mm   6. 52°8´   2. 75 cm   7. 290 mm^3

3. 48°12´   7. 47.68 km   3. 95.6 cm^2 8. 159 cm^3

4. 94.61 cm   8. 42.64 cm   4. 162 m^2 9. 201 m^2

5. 130 mm^2 10. 0.206 m^3

Part F   1. 310 m   2. 1°16´   3. 1100 m   4. 14.53 m